Phtp H Kahn
phkahn@hotmail.com

THE GREAT DERANGEMENT

The Randy L. and Melvin R. Berlin Family Lectures

The Great Derangement

CLIMATE CHANGE AND THE UNTHINKABLE

Amitav Ghosh

The University of Chicago Press

CHICAGO AND LONDON

The University of Chicago Press, Chicago 60637
The University of Chicago Press, Ltd., London
© 2016 by Amitav Ghosh
All rights reserved. Published 2016.
Printed in the United States of America

25 24 23 22 21 20 19 18 17   4 5

ISBN-13: 978-0-226-32303-9 (cloth)
ISBN-13: 978-0-226-32317-6 (e-book)

DOI: 10.7208/chicago/9780226323176.001.0001

The University of Chicago Press gratefully acknowledges the generous
support of the Randy L. and Melvin R. Berlin Family Endowment toward
the publication of this book.

Library of Congress Cataloging-in-Publication Data
Names: Ghosh, Amitav, 1956– author.
Title: The great derangement : climate change and the unthinkable /
    Amitav Ghosh.
Other titles: Randy L. and Melvin R. Berlin family lectures.
Description: Chicago ; London : The University of Chicago Press, 2016.
    | Series: The Randy L. and Melvin R. Berlin family lectures
Identifiers: LCCN 2016018232 | ISBN 9780226323039 (cloth : alk. paper)
    | ISBN 9780226323176 (e-book)
Subjects: LCSH: Climatic changes in literature.
Classification: LCC PN56.C612 G48 2016 | DDC 809/.9336—dc23 LC record
    available at https://lccn.loc.gov/2016018232

♾ This paper meets the requirements of ANSI/NISO Z39.48-1992
(Permanence of Paper).

FOR MUKUL KESAVAN

In memory of the 1978 tornado

CONTENTS

PART I

Stories

1.

Who can forget those moments when something that seems inanimate turns out to be vitally, even dangerously alive? As, for example, when an arabesque in the pattern of a carpet is revealed to be a dog's tail, which, if stepped upon, could lead to a nipped ankle? Or when we reach for an innocent looking vine and find it to be a worm or a snake? When a harmlessly drifting log turns out to be a crocodile?

It was a shock of this kind, I imagine, that the makers of *The Empire Strikes Back* had in mind when they conceived of the scene in which Han Solo lands the Millennium Falcon on what he takes to be an asteroid—but only to discover that he has entered the gullet of a sleeping space monster.

To recall that memorable scene now, more than thirty-five years after the making of the film, is to recognize its impossibility. For if ever there were a Han Solo, in the near or distant future, his assumptions about interplanetary objects are certain to be very different from those that prevailed in California at the time when the film was made. The humans of the future will surely understand, knowing what they presumably will know about the history of their forebears on Earth, that only in one, very brief era, lasting less than three centuries, did a significant number of their kind believe that planets and asteroids are inert.

2.

My ancestors were ecological refugees long before the term was invented.

They were from what is now Bangladesh, and their village was on the shore of the Padma River, one of the mightiest waterways in the land. The story, as my father told it, was this: one day in the mid-1850s the great river suddenly changed course, drowning the village; only a few of the inhabitants had managed to escape to higher ground. It was this catastrophe that had unmoored our forebears; in its wake they began to move westward and did not stop until the year 1856, when they settled once again on the banks of a river, the Ganges, in Bihar.

I first heard this story on a nostalgic family trip, as we were journeying down the Padma River in a steamboat. I was a child then, and as I looked into those swirling waters I imagined a great storm, with coconut palms bending over backward until their fronds lashed the ground; I envisioned women and children racing through howling winds as the waters rose behind them. I thought of my ancestors sitting huddled on an outcrop, looking on as their dwellings were washed away.

To this day, when I think of the circumstances that have shaped my life, I remember the elemental force that untethered my ancestors from their homeland and launched them on the series of journeys that preceded, and made possible, my own travels. When I look into my past the river seems to meet my eyes, staring back, as if to ask, Do you recognize me, wherever you are?

Recognition is famously a passage from ignorance to knowledge. To recognize, then, is not the same as an initial introduction. Nor does recognition require an exchange of words: more often than not we recognize mutely. And to recognize is by no means to understand that which meets the eye; comprehension need play no part in a moment of recognition.

The most important element of the word *recognition* thus lies in its first syllable, which harks back to something prior, an already existing awareness that makes possible the passage from ignorance to knowledge: a moment of recognition occurs when a prior awareness flashes before us, effecting an instant

4

change in our understanding of that which is beheld. Yet this flash cannot appear spontaneously; it cannot disclose itself except in the presence of its lost other. The knowledge that results from recognition, then, is not of the same kind as the discovery of something new: it arises rather from a renewed reckoning with a potentiality that lies within oneself.

This, I imagine, was what my forebears experienced on that day when the river rose up to claim their village: they awoke to the recognition of a presence that had molded their lives to the point where they had come to take it as much for granted as the air they breathed. But, of course, the air too can come to life with sudden and deadly violence—as it did in the Congo in 1988, when a great cloud of carbon dioxide burst forth from Lake Nyos and rolled into the surrounding villages, killing 1,700 people and an untold number of animals. But more often it does so with a quiet insistence—as the inhabitants of New Delhi and Beijing know all too well—when inflamed lungs and sinuses prove once again that there is no difference between the without and the within; between using and being used. These too are moments of recognition, in which it dawns on us that the energy that surrounds us, flowing under our feet and through wires in our walls, animating our vehicles and illuminating our rooms, is an all-encompassing presence that may have its own purposes about which we know nothing.

It was in this way that I too became aware of the urgent proximity of nonhuman presences, through instances of recognition that were forced upon me by my surroundings. I happened then to be writing about the Sundarbans, the great mangrove forest of the Bengal Delta, where the flow of water and silt is such that geological processes that usually unfold in deep time appear to occur at a speed where they can be followed from week to week and month to month. Overnight a stretch of riverbank will disappear, sometimes taking houses and people with it; but elsewhere a shallow mud bank will arise and within weeks the shore will have broadened by several feet. For the most part, these processes are of course cyclical. But even back then, in the first years of the twenty-first century, portents of accumulative and irreversible change could also be seen, in receding shorelines and a steady intrusion of salt water on lands that had previously been cultivated.

This is a landscape so dynamic that its very changeability leads to innumerable moments of recognition. I captured some of these in my notes from that time, as, for example, in these lines, written in May 2002: "I do believe it to be true that the land here is demonstrably alive; that it does not exist solely, or even incidentally, as a stage for the enactment of human history; that it is [itself] a protagonist." Elsewhere, in another note, I wrote, "Here even a child will begin a story about his grandmother with the words: 'in those days the river wasn't here and the village was not where it is . . .'"

Yet, I would not be able to speak of these encounters as instances of recognition if some prior awareness of what I was witnessing had not already been implanted in me, perhaps by childhood experiences, like that of going to look for my family's ancestral village; or by memories like that of a cyclone, in Dhaka, when a small fishpond, behind our walls, suddenly turned into a lake and came rushing into our house; or by my grandmother's stories of growing up beside a mighty river; or simply by the insistence with which the landscape of Bengal forces itself on the artists, writers, and filmmakers of the region.

But when it came to translating these perceptions into the medium of my imaginative life—into fiction, that is—I found myself confronting challenges of a wholly different order from those that I had dealt with in my earlier work. Back then, those challenges seemed to be particular to the book I was then writing, *The Hungry Tide*; but now, many years later, at a moment when the accelerating impacts of global warming have begun to threaten the very existence of low-lying areas like the Sundarbans, it seems to me that those problems have far wider implications. I have come to recognize that the challenges that climate change poses for the contemporary writer, although specific in some respects, are also products of something broader and older; that they derive ultimately from the grid of literary forms and conventions that came to shape the narrative imagination in precisely that period when the accumulation of carbon in the atmosphere was rewriting the destiny of the earth.

3.

That climate change casts a much smaller shadow within the landscape of literary fiction than it does even in the public arena is not hard to establish. To see that this is so, we need only glance through the pages of a few highly regarded literary journals and book reviews, for example, the *London Review of Books*, the *New York Review of Books*, the *Los Angeles Review of Books*, the *Literary Journal*, and the *New York Times Review of Books*. When the subject of climate change occurs in these publications, it is almost always in relation to nonfiction; novels and short stories are very rarely to be glimpsed within this horizon. Indeed, it could even be said that fiction that deals with climate change is almost by definition not of the kind that is taken seriously by serious literary journals: the mere mention of the subject is often enough to relegate a novel or a short story to the genre of science fiction. It is as though in the literary imagination climate change were somehow akin to extraterrestrials or interplanetary travel.

There is something confounding about this peculiar feedback loop. It is very difficult, surely, to imagine a conception of seriousness that is blind to potentially life-changing threats. And if the urgency of a subject were indeed a criterion of its seriousness, then, considering what climate change actually portends for the future of the earth, it should surely follow that this would be the principal preoccupation of writers the world over—and this, I think, is very far from being the case.

But why? Are the currents of global warming too wild to be navigated in the accustomed barques of narration? But the truth, as is now widely acknowledged, is that we have entered a time when the wild has become the norm: if certain literary forms are unable to negotiate these torrents, then they will have failed—and their failures will have to be counted as an aspect of the broader imaginative and cultural failure that lies at the heart of the climate crisis.

Clearly the problem does not arise out of a lack of information: there are surely very few writers today who are oblivious to the current disturbances in climate systems the world over. Yet, it is a striking fact that when novelists do choose to write about climate change it is almost always outside of fiction. A case in point is the work of Arundhati Roy: not only is she one of the finest prose stylists of our time, she is passionate and deeply informed about climate change. Yet all her writings on these subjects are in various forms of nonfiction.

Or consider the even more striking case of Paul Kingsnorth, author of *The Wake*, a much-admired historical novel set in eleventh-century England. Kingsnorth dedicated several years of his life to climate change activism before founding the influential Dark Mountain Project, "a network of writers, artists and thinkers who have stopped believing the stories our civilization tells itself." Although Kingsnorth has written a powerful nonfiction account of global resistance movements, as of the time of writing he has yet to publish a novel in which climate change plays a major part.

I too have been preoccupied with climate change for a long time, but it is true of my own work as well, that this subject figures only obliquely in my fiction. In thinking about the mismatch between my personal concerns and the content of my published work, I have come to be convinced that the discrepancy is not the result of personal predilections: it arises out of the peculiar forms of resistance that climate change presents to what is now regarded as serious fiction.

4.

In his seminal essay "The Climate of History," Dipesh Chakrabarty observes that historians will have to revise many of their fundamental assumptions and procedures in this era of the Anthropocene, in which "humans have become geological agents, changing the most basic physical processes of the earth." I would go further and add that the Anthropocene presents a challenge not only to the arts and humanities, but also to our commonsense understandings and beyond that to contemporary culture in general.

There can be no doubt, of course, that this challenge arises in part from the complexities of the technical language that serves as our primary window on climate change. But neither can there be any doubt that the challenge derives also from the practices and assumptions that guide the arts and humanities. To identify how this happens is, I think, a task of the utmost urgency: it may well be the key to understanding why contemporary culture finds it so hard to deal with climate change. Indeed, this is perhaps the most important question ever to confront *culture* in the broadest sense—for let us make no mistake: the climate crisis is also a crisis of culture, and thus of the imagination.

Culture generates desires—for vehicles and appliances, for

certain kinds of gardens and dwellings—that are among the principal drivers of the carbon economy. A speedy convertible excites us neither because of any love for metal and chrome, nor because of an abstract understanding of its engineering. It excites us because it evokes an image of a road arrowing through a pristine landscape; we think of freedom and the wind in our hair; we envision James Dean and Peter Fonda racing toward the horizon; we think also of Jack Kerouac and Vladimir Nabokov. When we see an advertisement that links a picture of a tropical island to the word *paradise*, the longings that are kindled in us have a chain of transmission that stretches back to Daniel Defoe and Jean-Jacques Rousseau: the flight that will transport us to the island is merely an ember in that fire. When we see a green lawn that has been watered with desalinated water, in Abu Dhabi or Southern California or some other environment where people had once been content to spend their water thriftily in nurturing a single vine or shrub, we are looking at an expression of a yearning that may have been midwifed by the novels of Jane Austen. The artifacts and commodities that are conjured up by these desires are, in a sense, at once expressions and concealments of the cultural matrix that brought them into being.

This culture is, of course, intimately linked with the wider histories of imperialism and capitalism that have shaped the world. But to know this is still to know very little about the specific ways in which the matrix interacts with different modes of cultural activity: poetry, art, architecture, theater, prose fiction, and so on. Throughout history these branches of culture have responded to war, ecological calamity, and crises of many sorts: why, then, should climate change prove so peculiarly resistant to their practices?

From this perspective, the questions that confront writers and artists today are not just those of the politics of the carbon economy; many of them have to do also with our own practices and the ways in which they make us complicit in the conceal-ments of the broader culture. For instance: if contemporary trends in architecture, even in this period of accelerating car-bon emissions, favor shiny, glass-and-metal-plated towers, do we not have to ask, What are the patterns of desire that are fed by these gestures? If I, as a novelist, choose to use brand names as elements in the depiction of character, do I not need to ask myself about the degree to which this makes me complicit in the manipulations of the marketplace?

In the same spirit, I think it also needs to be asked, What is it about climate change that the mention of it should lead to banishment from the preserves of serious fiction? And what does this tell us about culture writ large and its patterns of evasion?

In a substantially altered world, when sea-level rise has swallowed the Sundarbans and made cities like Kolkata, New York, and Bangkok uninhabitable, when readers and museum-goers turn to the art and literature of our time, will they not look, first and most urgently, for traces and portents of the altered world of their inheritance? And when they fail to find them, what should they—what can they—do other than to conclude that ours was a time when most forms of art and lit-erature were drawn into the modes of concealment that pre-vented people from recognizing the realities of their plight? Quite possibly, then, this era, which so congratulates itself on its self-awareness, will come to be known as the time of the Great Derangement.

5.

On the afternoon of March 17, 1978, the weather took an odd turn in north Delhi. Mid-march is usually a nice time of year in that part of India: the chill of winter is gone and the blazing heat of summer is yet to come; the sky is clear and the monsoon is far away. But that day dark clouds appeared suddenly and there were squalls of rain. Then followed an even bigger surprise: a hailstorm.

I was then studying for an MA at Delhi University while also working as a part-time journalist. When the hailstorm broke, I was in a library. I had planned to stay late, but the unseasonal weather led to a change of mind and I decided to leave. I was on my way back to my room when, on an impulse, I changed direction and dropped in on a friend. But the weather continued to worsen as we were chatting, so after a few minutes I decided to head straight back by a route that I rarely had occasion to take.

I had just passed a busy intersection called Maurice Nagar when I heard a rumbling sound somewhere above. Glancing over my shoulder I saw a gray, tube-like extrusion forming on the underside of a dark cloud: it grew rapidly as I watched, and then all of a sudden it turned and came whiplashing down to earth, heading in my direction.

Across the street lay a large administrative building. I sprinted over and headed toward what seemed to be an entrance. But the glass-fronted doors were shut, and a small crowd stood huddled outside, in the shelter of an overhang. There was no room for me there so I ran around to the front of the building. Spotting a small balcony, I jumped over the parapet and crouched on the floor.

The noise quickly rose to a frenzied pitch, and the wind began to tug fiercely at my clothes. Stealing a glance over the parapet, I saw, to my astonishment, that my surroundings had been darkened by a churning cloud of dust. In the dim glow that was shining down from above, I saw an extraordinary panoply of objects flying past—bicycles, scooters, lampposts, sheets of corrugated iron, even entire tea stalls. In that instant, gravity itself seemed to have been transformed into a wheel spinning upon the fingertip of some unknown power.

I buried my head in my arms and lay still. Moments later the noise died down and was replaced by an eerie silence. When at last I climbed out of the balcony, I was confronted by a scene of devastation such as I had never before beheld. Buses lay overturned, scooters sat perched on treetops, walls had been ripped out of buildings, exposing interiors in which ceiling fans had been twisted into tulip-like spirals. The place where I had first thought to take shelter, the glass-fronted doorway, had been reduced to a jumble of jagged debris. The panes had shattered, and many people had been wounded by the shards. I realized that I too would have been among the injured had I remained there. I walked away in a daze.

Long afterward, I am not sure exactly when or where, I hunted down the *Times of India*'s New Delhi edition of March 18. I still have the photocopies I made of it.

"30 Dead," says the banner headline, "700 Hurt As Cyclone Hits North Delhi."

Here are some excerpts from the accompanying report: "Delhi, March 17: At least 30 people were killed and 700 injured, many of them seriously, this evening when a freak funnel-shaped whirlwind, accompanied by rain, left in its wake death and devastation in Maurice Nagar, a part of Kingsway Camp, Roshanara Road and Kamla Nagar in the Capital. The injured were admitted to different hospitals in the Capital.

"The whirlwind followed almost a straight line. . . . Some eyewitnesses said the wind hit the Yamuna river and raised waves as high as 20 or 30 feet. . . . The Maurice Nagar road . . . presented a stark sight. It was littered with fallen poles . . . trees, branches, wires, bricks from the boundary walls of various institutions, tin roofs of staff quarters and dhabas and scores of scooters, buses and some cars. Not a tree was left standing on either side of the road."

The report quotes a witness: "I saw my own scooter, which I had abandoned on the road, during those terrifying moments, being carried away in the wind like a kite. We saw all this happening around but were dumbfounded. We saw people dying . . . but were unable to help them. The two tea-stalls at the Maurice Nagar corner were blown out of existence. At least 12 to 15 persons must have been buried under the debris at this spot. When the hellish fury had abated in just four minutes, we saw death and devastation around."

The vocabulary of the report is evidence of how unprecedented this disaster was. So unfamiliar was this phenomenon that the papers literally did not know what to call it: at a loss for words they resorted to "cyclone" and "funnel-shaped whirlwind."

Not till the next day was the right word found. The headlines of March 19 read, "A Very, Very Rare Phenomenon, Says Met Office": "It was a tornado that hit northern parts of the Capital yesterday—the first of its kind. . . . According to the Indian Meteorological Department, the tornado was about 50 metres wide and covered a distance of about five k.m. in the space of two or three minutes."

This was, in effect, the first tornado to hit Delhi—and indeed the entire region—in recorded meteorological history. And somehow I, who almost never took that road, who rarely visited that part of the university, had found myself in its path.

Only much later did I realize that the tornado's eye had passed directly over me. It seemed to me that there was something eerily apt about that metaphor: what had happened at that moment was strangely like a species of visual contact, of beholding and being beheld. And in that instant of contact something was planted deep in my mind, something irreducibly mysterious, something quite apart from the danger that I had been in and the destruction that I had witnessed; something that was not a property of the thing itself but of the manner in which it had intersected with my life.

6.

As is often the case with people who are waylaid by unpredictable events, for years afterward my mind kept returning to my encounter with the tornado. Why had I walked down a road that I almost never took, just before it was struck by a phenomenon that was without historical precedent? To think of it in terms of chance and coincidence seemed only to impoverish the experience: it was like trying to understand a poem by counting the words. I found myself reaching instead for the opposite end of the spectrum of meaning—for the extraordinary, the inexplicable, the confounding. Yet these too did not do justice to my memory of the event.

Novelists inevitably mine their own experience when they write. Unusual events being necessarily limited in number, it is but natural that these should be excavated over and again, in the hope of discovering a yet undiscovered vein.

No less than any other writer have I dug into my own past while writing fiction. By rights then, my encounter with the tornado should have been a mother lode, a gift to be mined to the last little nugget.

It is certainly true that storms, floods, and unusual weather events do recur in my books, and this may well be a legacy of the tornado. Yet oddly enough, no tornado has ever figured in my novels. Nor is this due to any lack of effort on my part. Indeed, the reason I still possess those cuttings from the *Times of India* is that I have returned to them often over the years, hoping to put them to use in a novel, but only to meet with failure at every attempt.

On the face of it there is no reason why such an event should be difficult to translate into fiction; after all, many novels are filled with strange happenings. Why then did I fail, despite my best efforts, to send a character down a road that is imminently to be struck by a tornado?

In reflecting on this, I find myself asking, What would I make of such a scene were I to come across it in a novel written by someone else? I suspect that my response would be one of incredulity; I would be inclined to think that the scene was a contrivance of last resort. Surely only a writer whose imaginative resources were utterly depleted would fall back on a situation of such extreme improbability?

Improbability is the key word here, so we have to ask, What does the word mean?

Improbable is not the opposite of *probable*, but rather an inflexion of it, a gradient in a continuum of probability. But what does probability—a mathematical idea—have to do with fiction?

The answer is: Everything. For, as Ian Hacking, a prominent historian of the concept, puts it, probability is a "manner of conceiving the world constituted without our being aware of it."

Probability and the modern novel are in fact twins, born at about the same time, among the same people, under a shared star that destined them to work as vessels for the containment of the same kind of experience. Before the birth of the modern novel, wherever stories were told, fiction delighted in the unheard-of and the unlikely. Narratives like those of *The Arabian Nights, The Journey to the West,* and *The Decameron* proceed by leaping blithely from one exceptional event to another. This, after all, is how storytelling must necessarily proceed, inasmuch as it is a recounting of "what happened"—for such an inquiry can arise only in relation to something out of the ordinary, which is but another way of saying "exceptional" or "unlikely." In essence, narrative proceeds by linking together moments and scenes that are in some way distinctive or different: these are, of course, nothing other than instances of exception.

Novels too proceed in this fashion, but what is distinctive about the form is precisely the concealment of those exceptional moments that serve as the motor of narrative. This is achieved through the insertion of what Franco Moretti, the literary theorist, calls "fillers." According to Moretti, "fillers function very much like the good manners so important in [Jane] Austen: they are both mechanisms designed to keep the 'narrativity' of life under control—to give a regularity, a 'style' to existence." It is through this mechanism that worlds are conjured up, through everyday details, which function "as *the opposite of narrative.*"

It is thus that the novel takes its modern form, through "the relocation of the unheard-of toward the background . . . while the everyday moves into the foreground."

Thus was the novel midwifed into existence around the world, through the banishing of the improbable and the insertion of the everyday. The process can be observed with exceptional clarity in the work of Bankim Chandra Chatterjee, a nineteenth-century Bengali writer and critic who self-consciously adopted the project of carving out a space in which realist European-style fiction could be written in the vernacular languages of India. Chatterjee's enterprise, undertaken in a context that was far removed from the metropolitan mainstream, is one of those instances in which a circumstance of exception reveals the true life of a regime of thought and practice.

Chatterjee was, in effect, seeking to supersede many old and very powerful forms of fiction, ranging from the ancient Indian epics to Buddhist Jataka stories and the immensely fecund Islamicate tradition of Urdu *dastaans*. Over time, these narrative forms had accumulated great weight and authority, extending far beyond the Indian subcontinent: his attempt to claim territory for a new kind of fiction was thus, in its own way, a heroic endeavor. That is why Chatterjee's explorations are of particular interest: his charting of this new territory puts the contrasts between the Western novel and other, older forms of narrative in ever-sharper relief.

In a long essay on Bengali literature, written in 1871, Chatterjee launched a frontal assault on writers who modeled their work on traditional forms of storytelling: his attack on this so-called Sanskrit school was focused precisely on the notion of "mere narrative." What he advocated instead was a style of writing that would accord primacy to "sketches of character and pictures of Bengali life."

What this meant, in practice, is very well illustrated by Chatterjee's first novel, *Rajmohan's Wife*, which was written in English in the early 1860s. Here is a passage: "The house of Mathur Ghose was a genuine specimen of mofussil [provincial] magnificence united with a mofussil want of cleanliness. . . . From the far-off paddy fields you could descry through the intervening foliage, its high palisades and blackened walls. On a nearer view might be seen pieces of plaster of a venerable antiquity prepared to bid farewell to their old and weather-beaten tenement."

Compare this with the following lines from Gustave Flaubert's *Madame Bovary*: "We leave the high road . . . whence the valley is seen. . . . The meadow stretches under a bulge of low hills to join at the back with the pasture land of the Bray country, while on the eastern side, the plain, gently rising, broadens out, showing as far as eye can follow its blond cornfields."

In both these passages, the reader is led into a "scene" through the eye and what it beholds: we are invited to "descry," to "view," to "see." In relation to other forms of narrative, this is indeed something new: instead of being told about what happened we learn about what was observed. Chatterjee has, in a sense, gone straight to the heart of the realist novel's "mimetic ambition": detailed descriptions of everyday life (or "fillers") are therefore central to his experiment with this new form.

Why should the rhetoric of the everyday appear at exactly the time when a regime of statistics, ruled by ideas of probability and improbability, was beginning to give new shapes to society? Why did fillers suddenly become so important? Moretti's answer is "'Because they *offer the kind of narrative pleasure compatible with the new regularity of bourgeois life*. Fillers turn the novel into a 'calm passion' . . . they are part of what Weber called the 'rationalization' of modern life: a process that begins in the economy and in the administration, but eventually pervades the sphere of free time, private life, entertainment, feelings. . . . Or in other words: fillers are an attempt at rationalizing the novelistic universe: turning it into a world of few surprises, fewer adventures, and no miracles at all."

This regime of thought imposed itself not only on the arts but also on the sciences. That is why *Time's Arrow, Time's Cycle*, Stephen Jay Gould's brilliant study of the geological theories of gradualism and catastrophism is, in essence, a study of narrative. In Gould's telling of the story, the catastrophist recounting of the earth's history is exemplified by Thomas Burnet's *Sacred Theory of the Earth* (1690) in which the narrative turns on events of "unrepeatable uniqueness." As opposed to this, the gradualist approach, championed by James Hutton (1726–97) and Charles Lyell (1797–1875), privileges slow processes that unfold over time at even, predictable rates. The central credo in this doctrine was "nothing could change otherwise than the way things were seen to change in the present." Or, to put it simply: "Nature does not make leaps."

The trouble, however, is that Nature does certainly jump, if not leap. The geological record bears witness to many fractures in time, some of which led to mass extinctions and the like: it was one such, in the form of the Chicxulub asteroid, that probably killed the dinosaurs. It is indisputable, in any event, that catastrophes waylay both the earth and its individual inhabitants at unpredictable intervals and in the most improbable ways.

Which, then, has primacy in the real world, predictable processes or unlikely events? Gould's response is "the only possible answer can be 'both and neither.'" Or, as the National Research Council of the United States puts it: "It is not known whether the relocation of materials on the surface of the Earth is dominated by the slower but continuous fluxes operating all the time or by the spectacular large fluxes that operate during short-lived cataclysmic events."

It was not until quite recently that geology reached this agnostic consensus. Through much of the era when geology—and also the modern novel—were coming of age, the gradualist (or "uniformitarian") view held absolute sway and catastrophism was exiled to the margins. Gradualists consolidated their victory by using one of modernity's most effective weapons: its insistence that it has rendered other forms of knowledge obsolete. So, as Gould so beautifully demonstrates, Lyell triumphed over his adversaries by accusing them of being primitive: "In an early stage of advancement, when a great number of natural appearances are unintelligible, an eclipse, an earthquake, a flood, or the approach of a comet, with many other occurrences afterwards found to belong to the regular course of events, are regarded as prodigies. The same delusion prevails as to moral phenomena, and many of these are ascribed to the intervention of demons, ghosts, witches, and other immaterial and supernatural agents."

This is exactly the rhetoric that Chatterjee uses in attacking the "Sanskrit school": he accuses those writers of depending on conventional modes of expression and fantastical forms of causality. "If love is to be the theme, Madana is invariably put into requisition with his five flower-tipped arrows; and the tyrannical king of Spring never fails to come to fight in his cause, with his army of bees, and soft breezes, and other ancient accompaniments. Are the pangs of separation to be sung? The moon is immediately cursed and anathematized, as scorching the poor victim with her cold beams."

Flaubert sounds a strikingly similar note in satirizing the narrative style that entrances the young Emma Rouault: in the novels that were smuggled into her convent, it was "all love, lovers, sweethearts, persecuted ladies fainting in lonely pavilions, postilions killed at every stage, horses ridden to death on every page, sombre forests, heartaches, vows, sobs, tears and kisses, little skiffs by moonlight, nightingales in shady groves." All of this is utterly foreign to the orderly bourgeois world that Emma Bovary is consigned to; such fantastical stuff belongs in the "dithyrambic lands" that she longs to inhabit.

In a striking summation of her tastes in narrative, Emma declares, "I . . . adore stories that rush breathlessly along, that frighten one. I detest commonplace heroes and moderate sentiments, such as there are in Nature."

"Commonplace"? "Moderate"? How did Nature ever come to be associated with words like these?

The incredulity that these associations evoke today is a sign of the degree to which the Anthropocene has already disrupted many assumptions that were founded on the relative climatic stability of the Holocene. From the reversed perspective of our time, the complacency and confidence of the emergent bourgeois order appears as yet another of those uncanny instances in which the planet seems to have been toying with humanity, by allowing it to assume that it was free to shape its own destiny.

Unlikely though it may seem today, the nineteenth century was indeed a time when it was assumed, in both fiction and geology, that Nature was moderate and orderly: this was a distinctive mark of a new and "modern" worldview. Chatterjee goes out of his way to berate his contemporary, the poet Michael Madhusudan Datta, for his immoderate portrayals of Nature: "Mr. Datta . . . wants repose. The winds rage their loudest when there is no necessity for the lightest puff. Clouds gather and pour down a deluge, when they need do nothing of the kind; and the sea grows terrible in its wrath, when everybody feels inclined to resent its interference."

The victory of gradualist views in science was similarly won by characterizing catastrophism as un-modern. In geology, the triumph of gradualist thinking was so complete that Alfred Wegener's theory of continental drift, which posited upheavals of sudden and unimaginable violence, was for decades discounted and derided.

It is worth recalling that these habits of mind held sway until late in the twentieth century, especially among the general public. "As of the mid-1960s," writes the historian John L. Brooke, "a gradualist model of earth history and evolution . . . reigned supreme." Even as late as 1985, the editorial page of the *New York Times* was inveighing against the asteroidal theory of dinosaur extinction: "Astronomers should leave to astrologers the task of seeking the causes of events in the stars." As for professional paleontologists, Elizabeth Kolbert notes, they reviled both the theory and its originators, Luis and Walter Alvarez: "'The Cretaceous extinctions were gradual and the catastrophe theory is wrong,' . . . [a] paleontologist stated. But 'simplistic theories will continue to come along to seduce a few scientists and enliven the covers of popular magazines.'"

In other words, gradualism became "a set of blinders" that eventually had to be put aside in favor of a view that recognizes the "twin requirements of uniqueness to mark moments of time as distinctive, and lawfulness to establish a basis of intelligibility."

Distinctive moments are no less important to modern novels than they are to any other forms of narrative, whether geological or historical. Ironically, this is nowhere more apparent than in *Rajmohan's Wife* and *Madame Bovary*, in both of which chance and happenstance are crucial to the narrative. In Flaubert's novel, for instance, the narrative pivots at a moment when Monsieur Bovary has an accidental encounter with his wife's soon-to-be lover at the opera, just after an impassioned scene during which she has imagined that the lead singer "was looking at her . . . She longed to run to his arms, to take refuge in his strength, as in the incarnation of love itself, and to say to him, to cry out, 'Take me away! carry me with you!'"

It could not, of course, be otherwise: if novels were not built upon a scaffolding of exceptional moments, writers would be faced with the Borgesian task of reproducing the world in its entirety. But the modern novel, unlike geology, has never been forced to confront the centrality of the improbable: the concealment of its scaffolding of events continues to be essential to its functioning. It is this that makes a certain kind of narrative a recognizably modern novel.

Here, then, is the irony of the "realist" novel: the very gestures with which it conjures up reality are actually a concealment of the real.

What this means in practice is that the calculus of probability that is deployed within the imaginary world of a novel is not the same as that which obtains outside it; this is why it is commonly said, "If this were in a novel, no one would believe it." Within the pages of a novel an event that is only slightly improbable in real life—say, an unexpected encounter with a long-lost childhood friend—may seem wildly unlikely: the writer will have to work hard to make it appear persuasive.

If that is true of a small fluke of chance, consider how much harder a writer would have to work to set up a scene that is wildly improbable even in real life? For example, a scene in which a character is walking down a road at the precise moment when it is hit by an unheard-of weather phenomenon?

To introduce such happenings into a novel is in fact to court eviction from the mansion in which serious fiction has long been in residence; it is to risk banishment to the humbler dwellings that surround the manor house—those generic outhouses that were once known by names such as "the Gothic," "the romance," or "the melodrama," and have now come to be called "fantasy," "horror," and "science fiction."

7.

So far as I know, climate change was not a factor in the tornado that struck Delhi in 1978. The only thing it has in common with the freakish weather events of today is its extreme improbability. And it appears that we are now in an era that will be defined precisely by events that appear, by our current standards of normalcy, highly improbable: flash floods, hundred-year storms, persistent droughts, spells of unprecedented heat, sudden landslides, raging torrents pouring down from breached glacial lakes, and, yes, freakish tornadoes.

The superstorm that struck New York in 2012, Hurricane Sandy, was one such highly improbable phenomenon: the word *unprecedented* has perhaps never figured so often in the description of a weather event. In his fine study of Hurricane Sandy, the meteorologist Adam Sobel notes that the track of the storm, as it crashed into the east coast of the United States, was without precedent: never before had a hurricane veered sharply westward in the mid-Atlantic. In turning, it also merged with a winter storm, thereby becoming a "mammoth hybrid" and attaining a size unprecedented in scientific memory. The storm surge that it unleashed reached a height that exceeded any in the region's recorded meteorological history.

Indeed, Sandy was an event of such a high degree of improbability that it confounded statistical weather-prediction models. Yet dynamic models, based on the laws of physics, were able to accurately predict its trajectory as well as its impacts.

But calculations of risk, on which officials base their decisions in emergencies, are based largely on probabilities. In the case of Sandy, as Sobel shows, the essential improbability of the phenomenon led them to underestimate the threat and thus delay emergency measures.

Sobel goes on to make the argument, as have many others, that human beings are intrinsically unable to prepare for rare events. But has this really been the case throughout human history? Or is it rather an aspect of the unconscious patterns of thought—or "common sense"—that gained ascendancy with a growing faith in "the regularity of bourgeois life"? I suspect that human beings were generally catastrophists at heart until their instinctive awareness of the earth's unpredictability was gradually supplanted by a belief in uniformitarianism—a regime of ideas that was supported by scientific theories like Lyell's, and also by a range of governmental practices that were informed by statistics and probability.

It is a fact, in any case, that when early tremors jolted the Italian town of L'Aquila, shortly before the great earthquake of 2009, many townsfolk obeyed the instinct that prompts people who live in earthquake-prone areas to move to open spaces. It was only because of a governmental intervention, intended to prevent panic, that they returned to their homes. As a result, a good number were trapped indoors when the earthquake occurred.

No such instinct was at work in New York during Sandy, where, as Sobel notes, it was generally believed that "losing one's life to a hurricane is . . . something that happens in faraway places" (he might just as well have said "dithyrambic lands"). In Brazil, similarly, when Hurricane Catarina struck the coast in 2004, many people did not take shelter because "they refused to believe that hurricanes were possible in Brazil."

But in the era of global warming, nothing is really far away; there is no place where the orderly expectations of bourgeois life hold unchallenged sway. It is as though our earth had become a literary critic and were laughing at Flaubert, Chatterjee, and their like, mocking *their* mockery of the "prodigious happenings" that occur so often in romances and epic poems.

This, then, is the first of the many ways in which the age of global warming defies both literary fiction and contemporary common sense: the weather events of this time have a very high degree of improbability. Indeed, it has even been proposed that this era should be named the "catastrophozoic" (others prefer such phrases as "the long emergency" and "the Penumbral Period"). It is certain in any case that these are not ordinary times: the events that mark them are not easily accommodated in the deliberately prosaic world of serious prose fiction.

Poetry, on the other hand, has long had an intimate relationship with climatic events: as Geoffrey Parker points out, John Milton began to compose *Paradise Lost* during a winter of extreme cold, and "unpredictable and unforgiving changes in the climate are central to his story. Milton's fictional world, like the real one in which he lived, was . . . a 'universe of death' at the mercy of extremes of heat and cold." This is a universe very different from that of the contemporary literary novel.

I am, of course, painting with a very broad brush: the novel's infancy is long past, and the form has changed in many ways over the last two centuries. Yet, to a quite remarkable degree, the literary novel has also remained true to the destiny that was charted for it at birth. Consider that the literary movements of the twentieth century were almost uniformly disdainful of plot and narrative; that an ever-greater emphasis was laid on style and "observation," whether it be of everyday details, traits of character, or nuances of emotion—which is why teachers of creative writing now exhort their students to "show, don't tell."

Yet fortunately, from time to time, there have also been movements that celebrated the unheard-of and the improbable: surrealism for instance, and most significantly, magical realism, which is replete with events that have no relation to the calculus of probability.

There is, however, an important difference between the weather events that we are now experiencing and those that occur in surrealist and magical realist novels: improbable though they might be, these events are neither surreal nor magical. To the contrary, these highly improbable occurrences are overwhelmingly, urgently, astoundingly real. The ethical difficulties that might arise in treating them as magical or metaphorical or allegorical are obvious perhaps. But there is another reason why, from the writer's point of view, it would serve no purpose to approach them in that way: because to treat them as magical or surreal would be to rob them of precisely the quality that makes them so urgently compelling—which is that they are actually happening on this earth, at this time.

8.

The Sundarbans are nothing like the forests that usually figure in literature. The greenery is dense, tangled, and low; the canopy is not above but around you, constantly clawing at your skin and your clothes. No breeze can enter the thickets of this forest; when the air stirs at all it is because of the buzzing of flies and other insects. Underfoot, instead of a carpet of softly decaying foliage, there is a bank of slippery, knee-deep mud, perforated by the sharp points that protrude from mangrove roots. Nor do any vistas present themselves except when you are on one of the hundreds of creeks and channels that wind through the landscape—and even then it is the water alone that opens itself; the forest withdraws behind its muddy ramparts, disclosing nothing.

In the Sundarbans, tigers are everywhere and nowhere. Often when you go ashore, you will find fresh tiger prints in the mud, but of the animal itself you will see nothing: glimpses of tigers are exceedingly uncommon and rarely more than fleeting. Yet you cannot doubt, since the prints are so fresh, that a tiger is somewhere nearby; and you know that it is probably watching you. In this jungle, concealment is so easy for an animal that it could be just a few feet away. If it charged, you would not see it till the last minute, and even if you did, you would not be able to get away; the mud would immobilize you.

Scattered through the forest are red rags, fluttering on branches. These mark the sites where people have been killed by tigers. There are many such killings every year; exactly how many no one knows because the statistics are not reliable. Nor is this anything new; in the nineteenth century, tens of thousands were killed by tigers. Suffice it to say that in some villages every household has lost a member to a tiger; everyone has a story to tell.

In these stories a great deal hinges on the eyes; seeing is one of their central themes; *not* seeing is another. The tiger is watching you; you are aware of its gaze, as you always are, but you do not see it; you do not lock eyes with it until it launches its charge, and at that moment a shock courses through you and you are immobilized, frozen.

The folk epic of the Sundarbans, *Bon Bibir Johuranama (The Miracles of Bon Bibi)*, comes to a climax in one such moment of mutual beholding, when the tiger demon, Dokkhin Rai, locks eyes with the protagonist, a boy called Dukhey.

It was then from afar, that the demon saw Dukhey . . .

Long had he hungered for this much-awaited prize; in an instant he assumed his tiger disguise.

"How long has it been since human flesh came my way? Now bliss awaits me in the shape of this boy Dukhey."

On the far mudbank Dukhey caught sight of the beast: "that tiger is the demon and I'm to be his feast."

Raising its head, the tiger reared its immense back; its jowls filled like sails as it sprang to attack.

The boy's life took wing, on seeing this fearsome sight.

Many stories of encounters with tigers hinge upon a moment of mutual recognition like this one. To look into the tiger's eyes is to recognize a presence of which you are already aware; and in that moment of contact you realize that this presence possesses a similar awareness of you, even though it is not human. This mute exchange of gazes is the only communication that is possible between you and this presence—yet communication it undoubtedly is.

But what is it that you are communicating with, at this moment of extreme danger, when your mind is in a state unlike any you've ever known before? An analogy that is sometimes offered is that of seeing a ghost, a presence that is not of this world.

In the tiger stories of the Sundarbans, as in my experience of the tornado, there is, as I noted earlier, an irreducible element of mystery. But what I am trying to suggest is perhaps better expressed by a different word, one that recurs frequently in translations of Freud and Heidegger. That word is *uncanny*— and it is indeed with uncanny accuracy that my experience of the tornado is evoked in the following passage: "In dread, as we say, 'one feels something uncanny.' What is this 'something' and this 'one'? We are unable to say what gives 'one' that uncanny feeling. 'One' just feels it generally."

It is surely no coincidence that the word *uncanny* has begun to be used, with ever greater frequency, in relation to climate change. Writing of the freakish events and objects of our era, Timothy Morton asks, "Isn't it the case, that the effect delivered to us in the [unaccustomed] rain, the weird cyclone, the oil slick is something uncanny?" George Marshall writes, "Climate change is inherently uncanny: Weather conditions, and the high-carbon lifestyles that are changing them, are extremely familiar and yet have now been given a new menace and uncertainty."

No other word comes close to expressing the strangeness of what is unfolding around us. For these changes are not merely strange in the sense of being unknown or alien; their uncanniness lies precisely in the fact that in these encounters we recognize something we had turned away from: that is to say, the presence and proximity of nonhuman interlocutors.

Yet now our gaze seems to be turning again; the uncanny and improbable events that are beating at our doors seem to have stirred a sense of recognition, an awareness that humans were never alone, that we have always been surrounded by beings of all sorts who share elements of that which we had thought to be most distinctively our own: the capacities of will, thought, and consciousness. How else do we account for the interest in the nonhuman that has been burgeoning in the humanities over the last decade and over a range of disciplines; how else do we account for the renewed attention to panpsychism and the metaphysics of Alfred North Whitehead; and for the rise to prominence of object-oriented ontology, actor-network theory, the new animism, and so on?

Can the timing of this renewed recognition be mere coincidence, or is the synchronicity an indication that there are entities in the world, like forests, that are fully capable of inserting themselves into our processes of thought? And if that were so, could it not *also* be said that the earth has itself intervened to revise those habits of thought that are based on the Cartesian dualism that arrogates all intelligence and agency to the human while denying them to every other kind of being?

This possibility is not, by any means, the most important of the many ways in which climate change challenges and refutes Enlightenment ideas. It is, however, certainly the most uncanny. For what it suggests—indeed proves—is that nonhuman forces have the ability to intervene directly in human thought. And to be alerted to such interventions is also to become uncannily aware that conversations among ourselves have always had other participants: it is like finding out that one's telephone has been tapped for years, or that the neighbors have long been eavesdropping on family discussions.

But in a way it's worse still, for it would seem that those unseen presences actually played a part in shaping our discussions without our being aware of it. And if these are real possibilities, can we help but suspect that all the time that we imagined ourselves to be thinking about apparently inanimate objects, we were ourselves being "thought" by other entities? It is almost as if the mind-altering planet that Stanislaw Lem

31

imagined in *Solaris* were our own, familiar Earth: what could be more uncanny than this?

These possibilities have many implications for the subject that primarily concerns me here, literary fiction. I will touch on some of these later, but for now I want to attend only to the aspect of the uncanny.

On the face of it, the novel as a form would seem to be a natural home for the uncanny. After all, have not some of the greatest novelists written uncanny tales? The ghost stories of Charles Dickens, Henry James, and Rabindranath Tagore come immediately to mind.

But the environmental uncanny is not the same as the uncanniness of the supernatural: it is different precisely because it pertains to nonhuman forces and beings. The ghosts of literary fiction are not human either, of course, but they are certainly represented as projections of humans who were once alive. But animals like the Sundarbans tiger, and freakish weather events like the Delhi tornado, have no human referents at all.

There is an additional element of the uncanny in events triggered by climate change, one that did not figure in my experience of the Delhi tornado. This is that the freakish weather events of today, despite their radically nonhuman nature, are nonetheless animated by cumulative human actions. In that sense, the events set in motion by global warming have a more intimate connection with humans than did the climatic phenomena of the past—this is because we have all contributed in some measure, great or small, to their making. They are the mysterious work of our own hands returning to haunt us in unthinkable shapes and forms.

All of this makes climate change events peculiarly resistant to the customary frames that literature has applied to "Nature": they are too powerful, too grotesque, too dangerous, and too accusatory to be written about in a lyrical, or elegiac, or romantic vein. Indeed, in that these events are not entirely of Nature (whatever that might be), they confound the very idea of "Nature writing" or ecological writing: they are instances, rather, of the uncanny intimacy of our relationship with the nonhuman.

More than a quarter century has passed since Bill McKibben wrote, "We live in a post-natural world." But did "Nature" in this sense ever exist? Or was it rather the deification of the human that gave it an illusory apartness from ourselves? Now that nonhuman agencies have dispelled that illusion, we are confronted suddenly with a new task: that of finding other ways in which to imagine the unthinkable beings and events of this era.

9.

In the final part of my novel *The Hungry Tide*, there is a scene in which a cyclone sends a gigantic storm surge into the Sundarbans. The wave results in the death of one of the principal characters, who gives his life protecting another.

This scene was extraordinarily difficult to write. In preparation for it, I combed through a great deal of material on catastrophic waves—storm surges as well as tsunamis. In the process, as often happens in writing fiction, the plight of the book's characters, as they faced the wave, became frighteningly real.

The Hungry Tide was published in the summer of 2004. A few months after the publication, on the night of December 25, I was back in my family home in Kolkata. The next morning, on logging on to the web, I learned that a cataclysmic tsunami had been set off by a massive undersea earthquake in the Indian Ocean. Measuring 9.0 on the Richter scale, the quake's epicenter lay between the northernmost tip of Sumatra and the southernmost island in the Andaman and Nicobar chain. Although the full extent of the catastrophe was not yet known, it was already clear that the toll in human lives would be immense.

The news had a deeply unsettling effect on me: the images that had been implanted in my mind by the writing of *The Hungry Tide* merged with live television footage of the tsunami in a way that was almost overwhelming. I became frantic; I could not focus on anything.

A couple of days later, I wrote to a newspaper and obtained a commission to write about the impact of the tsunami on the Andaman and Nicobar Islands. My first stop was the islands' capital city, Port Blair, which was thronged with refugees but had not suffered much damage itself: its location, above a sheltered cove, had protected it. After spending a few days there, I was able to board an Indian Air Force plane that was carrying supplies to one of the worst affected of the Nicobar Islands.

Unlike the Andamans, which rise steeply from the sea, the Nicobars are low-lying islands. Being situated close to the quake's epicenter, they had been very badly hit; many settlements had been razed. I visited a shoreside town called Malacca that had been reduced literally to its foundations: of the houses only the floors were left, and here and there the stump of a wall. It was as though the place had been hit by a bomb that was designed specifically to destroy all things human—for one of the strangest aspects of the scene was that the island's coconut palms were largely unaffected; they stood serenely amid the rubble, their fronds waving gently in the breeze that was blowing in from the sparkling, sun-drenched sea.

I wrote in my notebook: "The damage was limited to a half-mile radius along the shore. In the island's interior everything is tranquil, peaceful—indeed astonishingly beautiful. There are patches of tall, dark primary forest, beautiful padauk trees, and among these, in little clearings, huts built on stilts. . . . One of the ironies of the situation is that the most upwardly mobile people on the island were living at its edges."

Such was the pattern of settlement here that the indigenous islanders lived mainly in the interior: they were largely unaffected by the tsunami. Those who had settled along the seashore, on the other hand, were mainly people from the mainland, many of whom were educated and middle class: in settling where they had, they had silently expressed their belief that highly improbable events belong not in the real world but in fantasy. In other words, even here, in a place about as far removed as possible from the metropolitan centers that have shaped middle-class lifestyles, the pattern of settlement had come to reflect the uniformitarian expectations that are rooted in the "regularity of bourgeois life."

At the air force base where my plane had landed there was another, even more dramatic, illustration of this. The functional parts of the base—where the planes and machinery were kept—were located to the rear, well away from the water. The living areas, comprised of pretty little two-story houses, were built much closer to the sea, at the edge of a beautiful, palm-fringed beach. As always in military matters, the protocols of rank were strictly observed: the higher the rank of the officers, the closer their houses were to the water and the better the view that they and their families enjoyed.

Such was the design of the base that when the tsunami struck these houses the likelihood of survival was small, and inasmuch as it existed at all, it was in inverse relation to rank: the commander's house was thus the first to be hit.

The sight of the devastated houses was disturbing for reasons beyond the immediate tragedy of the tsunami and the many lives that had been lost there: the design of the base suggested a complacency that was itself a kind of madness. Nor could the siting of these buildings be attributed to the usual improvisatory muddle of Indian patterns of settlement. The base had to have been designed and built by a government agency; the site had clearly been chosen and approved by hardheaded military men and state appointed engineers. It was as if, in being adopted by the state, the bourgeois belief in the regularity of the world had been carried to the point of derangement.

A special place ought to be reserved in hell, I thought to myself, for planners who build with such reckless disregard for their surroundings.

Not long afterward, while flying into New York's John F. Kennedy airport, I looked out of the window and spotted Far Rockaway and Long Beach, the thickly populated Long Island neighborhoods that separate the airport from the Atlantic Ocean. Looking down on them from above, it was clear that those long rows of apartment blocks were sitting upon what had once been barrier islands, and that in the event of a major storm surge they would be swamped (as indeed they were when Hurricane Sandy hit the area in 2012). Yet it was clear also that these neighborhoods had not sprung up haphazardly; the sanction of the state was evident in the ordered geometry of their streets.

Only then did it strike me that the location of that base in the Nicobars was by no means anomalous; the builders had not in any sense departed from accepted global norms. To the contrary, they had merely followed the example of the European colonists who had founded cities like Bombay (Mumbai), Madras (Chennai), New York, Singapore, and Hong Kong, all of which are sited directly on the ocean. I understood also that what I had seen in the Nicobars was but a microcosmic expression of a pattern of settlement that is now dominant around the world: proximity to the water is a sign of affluence and education; a beachfront location is a status symbol; an ocean view greatly increases the value of real estate. A colonial vision of the world, in which proximity to the water represents power and security, mastery and conquest, has now been incorporated into the very foundations of middle-class patterns of living across the globe.

But haven't people always liked to live by the water?

Not really; through much of human history, people regarded the ocean with great wariness. Even when they made their living from the sea, through fishing or trade, they generally did not build large settlements on the water's edge: the great old port cities of Europe, like London, Amsterdam, Rotterdam, Stockholm, Lisbon, and Hamburg, are all protected from the open ocean by bays, estuaries, or deltaic river systems. The same is true of old Asian ports: Cochin, Surat, Tamluk, Dhaka, Mrauk-U, Guangzhou, Hangzhou, and Malacca are all cases in point. It is as if, before the early modern era, there had existed a general acceptance that provision had to be made for the unpredictable furies of the ocean—tsunamis, storm surges, and the like.

An element of that caution seems to have lingered even after the age of European global expansion began in the sixteenth century: it was not till the seventeenth century that colonial cities began to rise on seafronts around the world. Mumbai, Chennai, New York, and Charleston were all founded in this period. This would be followed by another, even more confident wave of city building in the nineteenth century, with the founding of Singapore and Hong Kong. These cities, all brought into being by processes of colonization, are now among those that are most directly threatened by climate change.

10.

Mumbai and New York, so different in so many ways, have in common that their destinies came to be linked to the British Empire at about the same time: the 1660s.

Although Giovanni da Verrazzano landed on Manhattan in 1524, the earliest European settlements in what is now New York State were built a long way up the Hudson River, in the area around Albany. It was not till 1625 that the Dutch built Fort Amsterdam on Manhattan island; this would later become New Amsterdam and then, when the British first seized the settlement in the 1660s, New York.

The site of today's Mumbai first came under European rule in 1535 when it was ceded to the Portuguese by the ruler of Gujarat. The site consisted of an estuarine archipelago, with a couple of large hilly islands to the north, close to the mainland, and a cluster of mainly low-lying islands to the south. This being an estuarine region, the relationship between land and water was so porous that the topography of the archipelago varied with the tides and the seasons.

Networks of shrines, villages, forts, harbors, and bazaars had existed on the southern islands for millennia, but they were never the site of an urban center as such, even in the early years of European occupation. The Portuguese built several churches and fortifications on those islands, but their main settlements were located close to the mainland, at Bassein, and on Salsette.

The southern part of the archipelago passed into British control when King Charles II married Catherine of Braganza in 1661: the islands were included in her dowry (which also contained a chest of tea: this was the Pandora's box that introduced the British public to the beverage, thereby setting in motion the vast cycles of trade that would turn nineteenth-century Bombay into the world's leading opium exporting port). It was only after passing into British hands that the southern islands became the nucleus of a sprawling urban conglomeration. It was then too that a distinct line of separation between land and sea was conjured up through the application of techniques of surveying within a "milieu of colonial power."

The appeal of the sites of both Mumbai and New York lay partly in their proximity to deepwater harbors and partly in the strategic advantages they presented: as islands, they were both easier to defend and easier to supply from the metropolis. A certain precariousness was thus etched upon them from the start by reason of their colonial origins.

The islands of south Mumbai did not long remain as they were when they were handed over to the British: links between them, in the form of causeways, bridges, embankments, and reclamation projects, began to rise in the eighteenth century. The reshaping of the estuarine landscape proceeded at such a pace that by the 1860s a Marathi chronicler, Govind Narayan, was able to predict with confidence that soon it would "never occur to anybody that Mumbai was an island once."

Today the part of the city that is located on the former islands to the south of Salsette has a population of about 11.8 million (the population of the Greater Mumbai area is somewhere in the region of 19 to 20 million). This promontory, less than twenty kilometers in length, is the center of many industries, including India's financial industry; the adjoining port handles more than half of the country's containerized cargo. This part of Mumbai is also home to many millionaires and billionaires: naturally many of them live along the western edge of the peninsula, which offers the finest views of the Arabian Sea.

Because of the density of its population and the importance of its institutions and industries, Mumbai represents an extraordinary, possibly unique, "concentration of risk." For this teeming metropolis, this great hub of economic, financial, and cultural activity, sits upon a wedge of cobbled-together land that is totally exposed to the ocean. It takes only a glance at a map to be aware of this: yet it was not till 2012 when Superstorm Sandy barreled into New York on October 29 that I began to think about the dangers of Mumbai's topography.

My wife and I were actually in Goa at the time, but since New York is also home to us we followed the storm closely, on the web and on TV, watching with mounting apprehension and disbelief as the storm swept over the city, devastating the oceanfront neighborhoods that we had flown over so many times while coming in to land at JFK airport.

As I watched these events unfold it occurred to me to wonder what would happen if a similar storm were to hit Mumbai. I reassured myself with the thought that this was very unlikely: both Mumbai and Goa face the Arabian Sea, which, unlike the Bay of Bengal, has not historically generated a great deal of cyclonic activity. Nor, unlike India's east coast, has the west coast had to deal with tsunamis: it was unaffected by the tsunami of 2004, for instance, which devastated large stretches of the eastern seaboard.

Still, the question intrigued me and I began to hunt for more information on the region's seismic and cyclonic profiles. Soon enough I learned that the west coast's good fortune might be merely a function of the providential protraction of geological time—for the Arabian Sea is by no means seismically inactive. A previously unknown, and probably very active, fault system was discovered in the Owen fracture zone a few years ago, off the coast of Oman; the system is eight hundred kilometers long and faces the west coast of India. This discovery was announced in an article that concludes with these words, chilling in their understatement: "These results will motivate a reappraisal of the seismic and tsunami hazard assessment in the NW Indian Ocean."

Soon, I also had to rethink my assumptions about cyclones and the Arabian Sea. Reading about Hurricane Sandy, I came upon more and more evidence that climate change may indeed alter patterns of cyclonic activity around the world: Adam So-

bel's *Storm Surge*, for example, suggests that significant changes
may be in the offing. When I began to look for information on
the Arabian Sea in particular, I learned that there had been an
uptick in cyclonic activity in those waters over the last couple
of decades. Between 1998 and 2001, three cyclones had crashed
into the Indian subcontinent to the north of Mumbai: they
claimed over seventeen thousand lives. Then in 2007, the Arabian Sea generated its strongest ever recorded storm: Cyclone
Gonu, a Category 5 hurricane, which hit Oman, Iran, and Pakistan in June that year causing widespread damage.

   What do these storms portend? Hoping to find an answer, I
reached out to Adam Sobel, who is a professor of atmospheric
science at Columbia University. He agreed to an interview, and
on a fine October day in 2015, I made my way to his Manhattan apartment. He confirmed to me that the most up-to-date
research indicates that the Arabian Sea is one of the regions of
the world where cyclonic activity is indeed likely to increase:
a 2012 paper by a Japanese research team predicts a 46 percent
increase in tropical cyclone frequency in the Arabian Sea by
the end of the next century, with a corresponding 31 percent
decrease in the Bay of Bengal. It also predicts another change:
in the past, cyclones were rare during the monsoon because
wind flows in the northern Indian Ocean were not conducive to
their formation in that season. Those patterns are now changing in such a way as to make cyclones more likely during and
after the monsoons. Another paper, by an American research
team, concludes that cyclonic activity in the Arabian Sea is
also likely to intensify because of the cloud of dust and pol-
lution that now hangs over the Indian subcontinent and its
surrounding waters: this too is contributing to changes in the
region's wind patterns.

   These findings prompted me to ask Adam whether he might
be willing to write a short piece assessing the risks that chang-

ing climatic patterns pose for Mumbai. He agreed and thus began a very interesting series of exchanges.

A few weeks after our meeting, Adam sent me this message:

> I have been doing a little research on Mumbai storm surge risk. There seems to be very little written about it. I have found a number of vague acknowledgments that the risk exists, but nothing that quantifies it.
>
> However, are you aware of the 1882 Mumbai cyclone? I have found only very brief accounts of it so far, but the death toll appears to have been between 100,000 and 200,000, and one source says there was a 6m storm surge, which is enormous, and I presume would account for much if not all of that! This was in one paragraph of a book that seems to be out of print. I haven't quickly found any more substantive sources online—most are single-line mentions in lists of deadly storms. I wonder if you have ever seen anything more in-depth?
>
> It is very spooky indeed that this storm is not mentioned in the various academic studies I have dug up on storm surge risk in India.

A quick Google search produced a number of references to an 1882 Bombay cyclone (some were even accompanied by pictures). There were several mentions of a death toll upward of one hundred thousand.

The figure astounded me. Mumbai's population then was about eight hundred thousand, which would mean that an eighth or more of the population would have perished: an extrapolation from these figures to today's Mumbai would yield a number of over a million.

But then came a surprise: Adam wrote to say that the 1882 cyclone was probably a hoax or rumor. He had not been able to find a reliable record of it; nor had any of the meteorologists or historians that he had written to. I then wrote to Murali Ranganathan, an expert on nineteenth-century Bombay, and he looked up the 1882 issues of the *Kaiser-i-Hind*, a Bombay-based Gujarati weekly run by Parsis. He found a brief description of a storm with strong winds and heavy rain on June 4, 1882, but there was no mention of any loss of life. Evidently, there was no great storm in 1882: it is a myth that has gained a life of its own.

However, the search did confirm that colonial Bombay had been struck by cyclones several times in the past; the 1909 edition of the city's *Gazetteer* even notes, "Since written history supplanted legend Bombay appears to have been visited somewhat frequently by great hurricanes and minor cyclonic storms."

Mumbai's earliest recorded encounter with a powerful storm was on May 15, 1618. A Jesuit historian described it thus: "The sky clouded, thunder burst, and a mighty wind arose. Towards nightfall a whirlwind raised the waves so high that the people, half dead from fear, thought that their city would be swallowed up. . . . The whole was like the ruin at the end of all things." Another Portuguese historian noted of this storm: "The sea was brought into the city by the wind; the waves roared fearfully; the tops of the churches were blown off and immense stones were driven to vast distances; two thousand persons were killed." If this figure is correct, it would suggest that the storm killed about a fifth of the population that then lived on the archipelago.

In 1740, another "terrific storm" caused great damage to the city, and in 1783 a storm that was "fatal to every ship in its path" killed four hundred people in Bombay harbor. The city was also hit by several cyclones in the nineteenth century: the worst was in 1854, when "property valued at half-a-million pounds sterling" was destroyed in four hours and a thousand people were killed.

Since the late nineteenth century onward, cyclones in the region seem to have "abated in number and intensity," but that may well be changing now. In 2009 Mumbai did experience a cyclonic storm, but fortunately its maximum wind speeds were in the region of 50 mph (85 kmph), well below those of a Category 1 hurricane on the Saffir-Simpson hurricane intensity scale. But encounters with storms of greater intensity may be forthcoming: 2015 was the first year in which the Arabian Sea is known to have generated more storms than the Bay of Bengal. This trend could tip the odds toward the recurrence of storms like those of centuries past.

Indeed, even as Adam and I were exchanging messages, Cyclone Chapala, a powerful storm, was forming in the Arabian Sea. Moving westward, it would hit the coast of Yemen on November 3, becoming the first Category 1 cyclone in recorded history to do so: in just two days, it would deluge the coast with more rain than it would normally get in several years. And then—as if to confirm the projections—even as Chapala was still battering Yemen, another cyclone, Megh, formed in the Arabian Sea and began to move along a similar track. A few days later another cyclone began to take shape in the Bay of Bengal, so that the Indian subcontinent was flanked by cyclones on both sides, a very rare event.

Suddenly the waters around India were churning with improbable events.

11.

What might happen if a Category 4 or 5 storm, with 150 mph or higher wind speeds, were to run directly into Mumbai? Mumbai's previous encounters with powerful cyclones occurred at a time when the city had considerably less than a million inhabitants; today it is the second-largest municipality in the world with a population of over 20 million. With the growth of the city, its built environment has also changed so that weather that is by no means exceptional often has severe effects: monsoon downpours, for instance, often lead to flooding nowadays. With an exceptional event the results can be catastrophic.

One such occurred on July 26, 2005, when a downpour without precedent in Mumbai's recorded history descended on the city: the northern suburbs received 94.4 cm of rain in fourteen hours, one of the highest rainfall totals ever recorded anywhere in a single day. On that day, with catastrophic suddenness, the people of the city were confronted with the costs of three centuries of interference with the ecology of an estuarine location.

The remaking of the landscape has so profoundly changed the area's topography that its natural drainage channels are now little more than filth-clogged ditches. The old waterways have been so extensively filled in, diverted, and built over that their carrying capacity has been severely diminished; and the water bodies, swamplands, and mangroves that might have served as natural sinks have also been encroached upon to a point where they have lost much of their absorptive ability.

A downpour as extreme as that of July 26 would pose a challenge even to a very effective drainage system: Mumbai's choked creeks and rivers were wholly inadequate to the onslaught. They quickly overflowed causing floods in which water was mixed with huge quantities of sewerage as well as dangerous industrial effluents. Roads and rail tracks disappeared under waist-high and even chest-high floodwaters; in the northern part of the city, where the rainfall was largely concentrated, entire neighborhoods were inundated: 2.5 million people "were under water for hours together."

On weekdays Mumbai's suburban railway network transports close to 6.6 million passengers; buses carry more than 1.5 million. The deluge came down on a Tuesday, beginning at around 2 p.m. Local train services were soon disrupted, and by 4:30 p.m. none were moving; several arterial roads and intersections were cut off by floodwaters at about the same time. The situation worsened as more and more vehicles poured on to the roads; in many parts of the city traffic came to a complete standstill. Altogether two hundred kilometers of road were submerged; some motorists drowned in their cars because short-circuited electrical systems would not allow them to open doors and windows. Thousands of scooters, motorcycles, cars, and buses were abandoned on the water-logged roads.

At around 5 p.m. cellular networks failed; most landlines stopped working too. Soon much of the city's power supply was also cut off (although not before several people had been electrocuted): parts of the city would remain without power for several days. Two million people, including many schoolchildren, were stranded, with no means of reaching home; a hundred and fifty thousand commuters were jammed into the city's two major railway stations. Those without money were unable to withdraw cash because ATM services had been knocked out as well.

Road, rail, and air services would remain cut off for two days. Over five hundred people died: many were washed away in the floods; some were killed in a landslide. Two thousand residential buildings were partially or completely destroyed; more than ninety thousand shops, schools, health care centers, and other buildings suffered damage.

While Mumbai's poor, especially the inhabitants of some of its informal settlements, were among the worst affected, the rich and famous were not spared either. The most powerful politician in the city had to be rescued from his home in a fishing boat; many Bollywood stars and industrialists were stranded or trapped by floodwaters.

Through all of this the people of Mumbai showed great generosity and resilience, sharing food and water and opening up their homes to strangers. Yet, as one observer notes, on July 26, 2005, it became "clear to many million people in Mumbai that life may never be quite the same again. An exceptional rainstorm finally put to rest the long prevailing myth of Mumbai's indestructible resilience to all kinds of shocks, including that of the partition."

In the aftermath of the deluge, many recommendations were made by civic bodies, NGOs, and even the courts. But ten years later, when another downpour occurred on June 10, 2015, it turned out that few of the recommended measures had been implemented: even though the volume of rainfall was only a third of that of the deluge of 2005, many parts of the city were again swamped by floodwaters.

What does Mumbai's experience of the downpour of 2005 tell us about what might, or might not, happen if a major storm happens to hit the city? The events will, of course, unfold very differently: to start with, a cyclone will arrive not with a few hours' notice, as was the case with the deluges, but after a warning period of several days. Storms are now so closely tracked, from the time they form onward, that there is usually an interval of a few days when emergency measures can be put in place.

Of these emergency measures, probably the most effective is evacuation. In historically cyclone-prone areas, like eastern India and Bangladesh, systems have been set up to move millions of people away from the coast when a major storm approaches; these measures have dramatically reduced casualties in recent years. But the uptick in cyclonic activity in the Arabian Sea is so recent that there has yet been no need for large-scale evacuations on the subcontinent's west coast. Whether such evacuations could be organized is an open question. Mumbai has been lucky not to have been hit by a major storm in more than a century; perhaps for that reason the possibility appears not to have been taken adequately into account in planning for disasters. Moreover, here, "as in most megacities, disaster management is focused on post-disaster response."

In Mumbai disaster planning seems to have been guided largely by concerns about events that occur with little or no warning, like earthquakes and deluges: evacuations usually follow rather than precede disasters of this kind. With a cyclone, given a lead-up period of several days, it would not be logistically impossible to evacuate large parts of the city before the storm's arrival: its rail and port facilities would certainly be able to move millions of people to safe locations on the mainland. But in order to succeed, such an evacuation would require years of planning and preparation; people in at-risk areas would also need to be educated about the dangers to which they might be exposed. And that exactly is the rub—for in Mumbai, as in Miami and many other coastal cities, these are often the very areas in which expensive new construction projects are located. Property values would almost certainly decline if residents were to be warned of possible risks—which is why builders and developers are sure to resist efforts to disseminate disaster-related information. One consequence of the last two decades of globalization is that real estate interests have acquired enormous power, not just in Mumbai but around the world; very few civic bodies, especially in the developing world, can hope to prevail against construction lobbies, even where it concerns public safety. The reality is that "growth" in many coastal cities around the world now depends on ensuring that a blind eye is turned toward risk.

Even with extensive planning and preparation the evacuation of a vast city is a formidable task, and not only for logistical reasons. The experience of New Orleans, in the days before Hurricane Katrina, or of New York before Sandy, or the city of Tacloban before Haiyan, tells us that despite the most dire warnings large numbers of people will stay behind; even mandatory evacuation orders will be disregarded by many. In the case of a megacity like Mumbai this means that hundreds of thousands, if not millions, will find themselves in harm's way when a cyclone makes landfall. Many will no doubt assume that having dealt with the floods of the recent past they will also be able to ride out a storm.

But the impact of a Category 4 or 5 cyclone will be very different from anything that Mumbai has experienced in living memory. During the deluges of 2005 and 2015 rain fell heavily on some parts of the city and lightly on others: the northern suburbs bore the brunt of the rainfall in both cases. The effects of the flooding were also most powerfully felt in low-lying areas and by the residents of ground-level houses and apartments; people living at higher elevations, and on the upper stories of tall buildings, were not as badly affected.

But the winds of a cyclone will spare neither low nor high; if anything, the blast will be felt most keenly by those at higher elevations. Many of Mumbai's tall buildings have large glass windows; few, if any, are reinforced. In a cyclone these exposed expanses of glass will have to withstand, not just hurricane-strength winds, but also flying debris. Many of the dwellings in Mumbai's informal settlements have roofs made of metal sheets and corrugated iron; cyclone-force winds will turn these, and the thousands of billboards that encrust the city, into deadly projectiles, hurling them with great force at the glass-wrapped towers that soar above the city.

Nor will a cyclone overlook those parts of the city that were spared the worst of the floods; to the contrary they will probably be hit first and hardest. The cyclones that have struck the west coast of India in the past have all traveled upward on a northeasterly tack, from the southern quadrant of the Arabian Sea. A cyclone moving in this direction would run straight into south Mumbai, where many essential civic and national institutions are located.

The southernmost tip of Mumbai consists of a tongue of low-lying land, much of it reclaimed; several important military and naval installations are located there, as is one of the country's most important scientific bodies—the Tata Institute of Fundamental Research. A storm surge of two or three meters would put much of this area under water; single-story buildings may be submerged almost to the roof. And an even higher surge is possible.

Not far from here lie the areas in which the city's most famous landmarks and institutions are located: most notably, the iconic Marine Drive, with its sea-facing hotels, famous for their sunset views, and its necklace-like row of art deco buildings. All of this sits on reclaimed land; at high tide waves often pour over the seawall. A storm surge would be barely impeded as it swept over and advanced eastward.

A distance of about four kilometers separates south Mumbai's two sea-facing shorelines. Situated on the east side are the city's port facilities, the legendary Taj Mahal Hotel, and the plaza of the Gateway of India, which is already increasingly prone to flooding. Beyond lies a much-used fishing port: any vessels that had not been moved to safe locations would be seized by the storm surge and swept toward the Gateway of India and the Taj Hotel.

At this point waves would be pouring into South Mumbai from both its sea-facing shorelines; it is not inconceivable that the two fronts of the storm surge would meet and merge. In that case the hills and promontories of south Mumbai would once again become islands, rising out of a wildly agitated expanse of water. Also visible above the waves would be the upper stories of many of the city's most important institutions: the Town Hall, the state legislature, the Chhatrapati Shivaji Railway Terminus, the towering headquarters of the Reserve Bank of India, and the skyscraper that houses India's largest and most important stock exchange.

Much of south Mumbai is low lying; even after the passing of the cyclone many neighborhoods would probably be waterlogged for several days; this will be true of other parts of the city as well. If the roads and rail lines are cut for any length of time, food and water shortages may develop, possibly leading to civil unrest. In Mumbai waterlogging often leads to the spread of illness and disease: the city's health infrastructure was intended to cater to a population of about half its present size; its municipal hospitals have only forty thousand beds. Since many hospitals will have been evacuated before the storm, it may be difficult for the sick and injured to get medical attention. If Mumbai's stock exchange and Reserve Bank are rendered inoperative, then India's financial and commercial systems may be paralyzed.

But there is another possibility, yet more frightening. Of the world's megacities, Mumbai is one of the few that has a nuclear facility within its urban limits: the Bhabha Atomic Research Centre at Trombay. To the north, at Tarapur, ninety-four kilometers from the city's periphery, lies another nuclear facility. Both these plants sit right upon the shoreline, as do many other nuclear installations around the world: these locations were chosen in order to give them easy access to water.

With climate change many nuclear plants around the world are now threatened by rising seas. An article in the *Bulletin of the Atomic Scientists* notes: "During massive storms . . . there is a greatly increased chance of the loss of power at a nuclear power plant, which significantly contributes to safety risks." Essential cooling systems could fail; safety systems could be damaged; contaminants could seep into the plant and radioactive water could leak out, as happened at the Fukushima Daiichi plant.

What threats might a major storm pose for nuclear plants like those in Mumbai's vicinity? I addressed this question to a nuclear safety expert, M. V. Ramana, of the Program on Science and Global Security at Princeton University. His answer was as follows: "My biggest concerns have to do with the tanks in which liquid radioactive waste is stored. These tanks contain, in high concentrations, radioactive fission products and produce a lot of heat due to radioactive decay; explosive chemicals can also be produced in these tanks, in particular hydrogen gas. Typically waste storage facilities include several safety systems to prevent explosions. During major storms, however, some or all of these systems could be simultaneously disabled: cascading failures could make it difficult for workers to carry out any repairs—this is assuming that there will be any workers available and capable of undertaking repairs during a major storm. An explosion at such a tank, depending on the energy of the explosion and the exact weather conditions, could lead to the dispersal of radioactivity over hundreds of square kilometers; this in turn could require mass evacuations or the long-term cessation of agriculture in regions of high contamination."

Fortunately, the chances of a cyclone hitting Mumbai are small in any given year. But there is no doubt whatsoever of the threats that will confront the city because of other climate change impacts: increased precipitation and rising sea levels. If there are substantial increases in rainfall over the next few decades, as climate models predict, then damaging floods will become more frequent. As for sea levels, if they rise by a meter or more by the end of the century, as some climate scientists fear they might, then some parts of south Mumbai will gradually become uninhabitable.

A similar fate awaits two other colonial cities, founded in the same century as Mumbai: Chennai (Madras), which also experienced a traumatic deluge in 2015; and Kolkata, to which I have close familial links.

Unlike Chennai and Mumbai, Kolkata is not situated beside the sea. However, much of its surface area is below sea level, and the city is subject to regular flooding: like everyone who has lived in Kolkata, I have vivid memories of epic floods. But long familiarity with flooding tends to have a lulling effect, which is why it came as a shock to me when I learned, from a World Bank report, that Kolkata is one of the global megacities that is most at risk from climate change; equally shocking was the discovery that my family's house, where my mother and sister live, is right next to one of the city's most threatened neighborhoods.

The report forced me to face a question that eventually confronts everybody who takes the trouble to inform themselves about climate change: what can I do to protect my family and loved ones now that I know what lies ahead? My mother is elderly and increasingly frail; there is no telling how she would fare if the house were to be cut off by a flood and medical attention were to become unavailable for any length of time.

After much thought I decided to talk to my mother about moving. I tried to introduce the subject tactfully, but it made little difference: she looked at me as though I had lost my mind. Nor could I blame her: it *did* seem like lunacy to talk about leaving a beloved family home, with all its memories and associations, simply because of a threat outlined in a World Bank report.

It was a fine day, cool and sunlit; I dropped the subject.

But the experience did make me recognize something that I

would otherwise have been loathe to admit: contrary to what I might like to think, my life is not guided by reason; it is ruled, rather, by the inertia of habitual motion. This is indeed the condition of the vast majority of human beings, which is why very few of us will be able to adapt to global warming if it is left to us, as individuals, to make the necessary changes; those who will uproot themselves and make the right preparations are precisely those obsessed monomaniacs who appear to be on the borderline of lunacy.

If whole societies and polities are to adapt then the necessary decisions will need to be made collectively, within political institutions, as happens in wartime or national emergencies. After all, isn't that what politics, in its most fundamental form, is about? Collective survival and the preservation of the body politic?

Yet, to look around the world today is to recognize that with some notable exceptions, like Holland and China, there exist very few polities or public institutions that are capable of implementing, or even contemplating, a managed retreat from vulnerable locations. For most governments and politicians, as for most of us as individuals, to leave the places that are linked to our memories and attachments, to abandon the homes that have given our lives roots, stability, and meaning, is nothing short of unthinkable.

12.

It is surely no accident that colonial cities like Mumbai, New York, Boston, and Kolkata were all brought into being through early globalization. They were linked to each other not only through the circumstances of their founding but also through patterns of trade that expanded and accelerated Western economies. These cities were thus the drivers of the very processes that now threaten them with destruction. In that sense, their predicament is but an especially heightened instance of a plight that is now universal.

It isn't only in retrospect that the siting of some of these cities now appear as acts of utter recklessness: Bombay's first Parsi residents were reluctant to leave older, more sheltered ports like Surat and Navsari and had to be offered financial incentives to move to the newly founded city. Similarly, Qing dynasty officials were astonished to learn that the British intended to build a city on the island of Hong Kong: why would anyone want to create a settlement in a place that was so exposed to the vagaries of the earth?

But in time, sure enough, there was a collective setting aside of the knowledge that accrues over generations through dwelling in a landscape. People began to move closer and closer to the water.

How did this come about? The same question arises also in relation to the coast around Fukushima, where stone tablets had been placed along the shoreline in the Middle Ages to serve as tsunami warnings; future generations were explicitly told "Do not build your homes below this point!"

The Japanese are certainly no more inattentive to the words of their ancestors than any other people: yet not only did they build *exactly* where they had been warned *not* to, they actually situated a nuclear plant there.

This too is an aspect of the uncanny in the history of our relations with our environments. It is not as if we had not been warned; it is not as if we were ignorant of the risks. An awareness of the precariousness of human existence is to be found in every culture: it is reflected in biblical and Quranic images of the Apocalypse, in the figuring of the Fimbulwinter in Norse mythology, in tales of *pralaya* in Sanskrit literature, and so on. It was the literary imagination, most of all, that was everywhere informed by this awareness.

Why then did these intuitions withdraw, not just from the minds of the founders of colonial cities, but also from the forefront of the literary imagination? Even in the West, the earth did not come to be regarded as moderate and orderly until long after the advent of modernity: for poets and writers, it was not until the late nineteenth century that Nature lost the power to evoke that form of terror and awe that was associated with the "sublime." But the practical men who ran colonies and founded cities had evidently acquired their indifference to the destructive powers of the earth much earlier.

How did this come about? How did a state of consciousness come into being such that millions of people would move to such dangerously exposed locations?

The chronology of the founding of these cities creates an almost irresistible temptation to point to the European Enlightenment's predatory hubris in relation to the earth and its resources. But this would tell us very little about the thinking of the men who built and planned that base in the Nicobars: if hubris and predation had anything to do with their choice of site, it was at a great remove. Between them and the cartographers and surveyors of an earlier era there was, I think, a much more immediate link: a habit of mind that proceeded by creating discontinuities; that is to say, they were trained to break problems into smaller and smaller puzzles until a solution presented itself. This is a way of thinking that deliberately excludes things and forces ("externalities") that lie beyond the horizon of the matter at hand: it is a perspective that renders the interconnectedness of Gaia unthinkable.

The urban history of Bengal provides an interesting illustration of what I am trying to get at. Colonial Calcutta, which was for a long time the capital of the British Raj, was founded on the banks of the Hooghly River in the late seventeenth century. It had not been in existence for long before it came to be realized that the river was silting up. By the early nineteenth century, the East India Company had decided in principle that a new port would be built at a location closer to the Bay of Bengal. A site was chosen in the 1840s; it lay some thirty-five miles to the south east of Calcutta, on the banks of a river called Matla (which means "crazed" or "intoxicated" in Bengali).

At that time, there lived in Calcutta an Englishman by the name of Henry Piddington. A shipping inspector by profession, he dabbled promiscuously in literature, philology, and the sciences until his true calling was revealed to him by a treatise, *An Attempt to Develop the Law of Storms*, written by an American meteorologist, Col. Henry Reid. Published in 1838, the book was an ambitious study of the circular motion of tropical storms. Colonel Reid's book inspired a great passion in Piddington, and he devoted himself to the field for the rest of his life. It was he who coined the word *cyclone*, and it is for this that he is best remembered today. But Piddington's particular interest was the phenomenon of the storm surge (or "storm wave" as it was then called): he would eventually compile a detailed account of storm surges along the coast of Bengal and the devastation they had caused.

Because of his familiarity with this subject, Piddington understood that the proposed port on the Matla River would be exposed to extreme cyclonic hazard. Such was his alarm that in 1853 he published a pamphlet, addressed to the then governor-general, in which he issued this ominous warning: "every one and everything must be prepared to see a day when, in the midst of the horrors of a hurricane, they will find a terrific mass of salt water rolling in, or rising up upon them, with such rapidity that the whole settlement will be inundated to a depth from five to fifteen feet."

Piddington's warnings fell on deaf ears: to the builders and civil servants who were working on the new city, he must have sounded like a madman—in the measureable, discrete universes that they worked within there was no place for a phenomenon that took birth hundreds of miles away and came storming over the seas like a "wonderful meteor" (to use Piddington's words).

It was probably the very scale of the phenomenon invoked by Piddington that made it unthinkable to those eminently practical men, accustomed as they were to the "regularity of bourgeois life." The port continued to rise, even through the great uprising of 1857: it was built on a lavish scale, with banks, hotels, a railway station, and imposing public buildings. The city was formally inaugurated in 1864 with a grand ceremony: it was named Port Canning, after a former governor-general.

Port Canning's claims to grandeur were short-lived. A mere three years after its inauguration, it was struck by a cyclone, just as Piddington had predicted. And even though the accompanying storm surge was a modest one, rising to only six feet, it caused terrible destruction. The city was abandoned four years later (Canning is now a small river port and access point for the Sundarbans). Piddington thus became one of the first Cassandras of climate science.

13.

If I have dwelt on this at some length, it is because the discontinuities that I have pointed to here have a bearing also on the ways in which worlds are created within novels. A "setting" is what allows most stories to unfold; its relation to the action is as close as that of a stage to a play. When we read *Middlemarch* or *Buddenbrooks* or *Waterland*, or the great Bengali novel *A River*

Called Titash, we enter into their settings until they begin to seem real to us; we ourselves become emplaced within them. This exactly is why "a sense of place" is famously one of the great conjurations of the novel as a form.

What the settings of fiction have in common with sites measured by surveyors is that they too are constructed out of discontinuities. Since each setting is particular to itself, its connections to the world beyond are inevitably made to recede (as, for example, with the imperial networks that make possible the worlds portrayed by Jane Austen and Charlotte Brontë). Unlike epics, novels do not usually bring multiple universes into conjunction; nor are their settings transportable outside their context in the manner of, say, the Ithaca of the *Odyssey* or the Ayodhya of the *Ramayana*.

In fiction, the immediate discontinuities of place are nested within others: Maycomb, Alabama, the setting of *To Kill a Mockingbird*, becomes a stand-in for the whole of the Deep South; the *Pequod*, the small Nantucket whaling vessel in *Moby Dick*, becomes a metaphor for America. In this way, settings become the vessel for the exploration of that ultimate instance of discontinuity: the nation-state.

In novels discontinuities of space are accompanied also by discontinuities of time: a setting usually requires a "period"; it is actualized within a certain time horizon. Unlike epics, which often range over eons and epochs, novels rarely extend beyond a few generations. The *longue durée* is not the territory of the novel.

It is through the imposition of these boundaries, in time and space, that the world of a novel is created: like the margins of a page, these borders render places into texts, so that they can be read. The process is beautifully illustrated in the opening pages of *A River Called Titash*. Published in 1956, this remarkable novel was the only work of fiction published by

Adwaita Mallabarman, who belonged to an impoverished caste of Dalit fisherfolk. The novel is set in rural Bengal, in a village on the shores of a fictional river, Titash.

Bengal is, as I have said, a land of titanic rivers. Mallabarman gestures toward the vastness of the landscape with these words: "The bosom of Bengal is draped with rivers and their tributaries, twisted and intertwined like tangled locks, streaked with the white of foamy waves."

But almost at once he begins to detach the setting of his novel from the larger landscape: all rivers are not the same, he tells us, some are like "a frenzied sculptor at work, destroying and creating restlessly in crazed joy, riding the high-flying swing of fearsome energy—here is one kind of art."

And then, in a striking passage, the writer announces his own intentions and premises: "There is another kind of art. . . . The practitioner of this art cannot depict Mahakaal (Shiva the Destroyer) in his cosmic dance of creation and destruction— the awesome vision of tangled brown hair tumbling out of the coiled mass will not come from this artist's brush. The artist has come away from the rivers Padma, Meghna, and Dhaleswari to find a home beside Titash.

"The pictures this artist draws please the heart. Little villages dot the edges of the water. Behind these villages are stretches of farmland."

This is how the reader learns that the Titash, although it is a part of a landscape of immense waterways, is itself a small, relatively gentle river: "No cities or large towns ever grew up on its banks. Merchant boats with giant sails do not travel its waters. Its name is not in the pages of geography books."

In this way, through a series of successive exclusions, Mallabarman creates a space that will submit to the techniques of a modern novel: the rest of the landscape is pushed farther and farther into the background until at last we have a setting that can carry a narrative. The setting becomes, in a sense, a self-contained ecosystem, with the river as the sustainer both of life and of the narrative. The impetus of the novel, and its poignancy, come from the Titash itself: it is the river's slow drying up that directs the lives of the characters. The Titash is, of course, but one strand of the "tangled locks" of an immense network of rivers and its flow is necessarily ruled by the dynamics of the landscape. But it is precisely by excluding those inconceivably large forces, and by telescoping the changes into the duration of a limited-time horizon, that the novel becomes narratable.

Contrast this with the universes of boundless time and space that are conjured up by other forms of prose narrative. Here, for example, are a couple of passages from the beginning of the sixteenth-century Chinese folk epic *The Journey to the West*: "At this point the firmament first acquired its foundation. With another 5,400 years came the Tzu epoch; the ethereal and the light rose up to form the four phenomena of the sun, the moon, the stars, and the heavenly bodies. . . . Following P'an Ku's construction of the universe . . . the world was divided into four great continents. . . . Beyond the ocean there was a country named Ao-lai. It was near a great ocean, in the midst of which was located the famous Flower-Fruit Mountain."

Here is a form of prose narrative, still immensely popular, that ranges widely and freely over vast expanses of time and space. It embraces the inconceivably large almost to the same degree that the novel shuns it. Novels, on the other hand, conjure up worlds that become real precisely because of their finitude and distinctiveness. Within the mansion of serious fiction, no one will speak of how the continents were created; nor will they refer to the passage of thousands of years: connections and events on this scale appear not just unlikely but also absurd within the delimited horizon of a novel—when they intrude, the temptation to lapse into satire, as in Ian McE-wan's *Solar*, becomes almost irresistible.

But the earth of the Anthropocene is precisely a world of insistent, inescapable continuities, animated by forces that are nothing if not inconceivably vast. The waters that are invading the Sundarbans are also swamping Miami Beach; deserts are advancing in China as well as Peru; wildfires are intensifying in Australia as well as Texas and Canada.

There was never a time, of course, when the forces of weath-er and geology did *not* have a bearing on our lives—but neither has there ever been a time when they have pressed themselves on us with such relentless directness. We have entered, as Tim-othy Morton says, the age of hyperobjects, which are defined in part by their stickiness, their ever-firmer adherence to our lives: even to speak of the weather, that safest of subjects, is now to risk a quarrel with a denialist neighbor. No less than they mock the discontinuities and boundaries of the nation-state do these connections defy the boundedness of "place," creating continuities of experience between Bengal and Loui-siana, New York and Mumbai, Tibet and Alaska.

I was recently sent a piece about a mangrove forest in Pap-ua New Guinea. This was once a "place" in the deepest sense that it was linked to its inhabitants through a dense web of mutual sustenance and symbolism. But in the wet season of 2007, "the barrier beaches were breached, cutting innumerable channels through to the lakes. Sand poured through them. Tidal surges tore across the villages, leaving behind a spectacle of severed trunks of coconut palms and dead shoreline trees, drifting canoes, trenches, and gullies. Entire villages had to be evacuated." Eventually the inhabitants were forced to abandon their villages.

The Anthropocene has reversed the temporal order of mo-dernity: those at the margins are now the first to experience the future that awaits all of us; it is they who confront most directly what Thoreau called "vast, Titanic, inhuman nature." Nor is it any longer possible to exclude this dynamic even from places that were once renowned for their distinctiveness. Can anyone write about Venice any more without mentioning the *aqua alta*, when the waters of the lagoon swamp the city's streets and courtyards? Nor can they ignore the relationship that this has with the fact that one of the languages most frequently heard in Venice is Bengali: the men who run the quaint little vegetable stalls and bake the pizzas and even play the accordion are largely Bangladeshi, many of them displaced by the same phenomenon that now threatens their adopted city—sea-level rise.

Behind all of this lie those continuities and those inconceivably vast forces that have now become impossible to exclude, even from texts.

Here, then, is another form of resistance, a scalar one, that the Anthropocene presents to the techniques that are most closely identified with the novel: its essence consists of phenomena that were long ago expelled from the territory of the novel—forces of unthinkable magnitude that create unbearably intimate connections over vast gaps in time and space.

14.

I would like to return, for a moment, to the images I started with: of apparently inanimate things coming suddenly alive. This, as I said earlier, is one of the uncanniest effects of the Anthropocene, this renewed awareness of the elements of agency and consciousness that humans share with many other beings, and even perhaps the planet itself.

But such truth as this statement has is only partial: for the fact is that a great number of human beings had never lost this awareness in the first place. In the Sundarbans, for example, the people who live in and around the mangrove forest have never doubted that tigers and many other animals possess intelligence and agency. For the first peoples of the Yukon, even glaciers are endowed with moods and feelings, likes and dislikes. Nor would these conceptions have been unthinkable for a scientist like Sir Jagadish Chandra Bose, who attributed elements of consciousness to vegetables and even metals, or for the primatologist Imanishi Kinji who insisted on "the unity of all elements on the planet earth—living and non-living."

Neither is it the case that we were all equally captive to Cartesian dualism before the awareness of climate change dawned on us: my ancestors were certainly not in its thrall, and even I was never fully acculturated to that view of the world. Indeed, I would venture to say that this is true for most people in the world, even in the West. To the great majority of people everywhere, it has always been perfectly evident that dogs, horses, elephants, chimpanzees, and many other animals possess intelligence and emotions. Did anyone ever really believe, *pace* Descartes, that animals are automatons? "Surely Descartes never saw an ape" wrote Linnaeus, who found it no easy matter to draw a line between human and animal. Even the most devoted Cartesian will probably have no difficulty in interpreting the emotions of a dog that has backed him up against a wall.

Nowhere is the awareness of nonhuman agency more evident than in traditions of narrative. In the Indian epics—and this is a tradition that remains vibrantly alive to this day— there is a completely matter-of-fact acceptance of the agency of nonhuman beings of many kinds. I refer not only to systems of belief but also to techniques of storytelling: nonhumans provide much of the momentum of the epics; they create the resolutions that allow the narrative to move forward. In the *Iliad* and the *Odyssey* too the intervention of gods, animals, and the elements is essential to the machinery of narration. This is true for many other narrative traditions as well, Asian, African, Mediterranean, and so on. The Hebrew Bible is no exception; as the theologian Michael Northcott points out, "At the heart of Judaism is a God who is encountered through Nature and events rather than words or texts. Christianity, by contrast, and then Islam, is a form of religion that is less implicated in the weather, climate and political power and more invested in words and texts."

But even within Christianity, it was not till the advent of Protestantism perhaps that Man began to dream of achieving his own self-deification by radically isolating himself before an arbitrary God. Yet that dream of silencing the nonhuman has never been completely realized, not even within the very heart of contemporary modernity; indeed, it would seem that one aspect of the agency of nonhumans is their uncanny ability to stay abreast of technology. Even among today's teenagers and twenty-somethings, whose most intimate familiars are man-made objects like iPads and iPhones, an awareness evidently still lingers that elements of agency are concealed everywhere within our surroundings: why else should the charts of best-selling books and top-grossing films continue to be so heavily weighted in favor of those that feature werewolves, vampires, witches, shape-shifters, extraterrestrials, mutants, and zombies?

So the real mystery in relation to the agency of nonhumans lies not in the renewed recognition of it, but rather in how this awareness came to be suppressed in the first place, at least within the modes of thought and expression that have become dominant over the last couple of centuries. Literary forms have clearly played an important, perhaps critical, part in the process. So, if for a moment, we were to take seriously the premise that I started with—that the Anthropocene has forced us to recognize that there are other, fully aware eyes looking over our shoulders—then the first question to present itself is this: What is the place of the nonhuman in the modern novel?

To attempt an answer is to confront another of the uncanny effects of the Anthropocene: it was in exactly the period in which human activity was changing the earth's atmosphere that the literary imagination became radically centered on the human. Inasmuch as the nonhuman was written about at all, it was not within the mansion of serious fiction but rather in the outhouses to which science fiction and fantasy had been banished.

15.

The separation of science fiction from the literary mainstream came about not as the result of a sudden drawing of boundaries but rather through a slow and gradual process. There was, however, one moment that was critical to this process, and it happens to have had a link to a climate-related event.

The seismic event that began on April 5, 1815, on Mount Tambora, three hundred kilometers to the east of Bali, was the greatest volcanic eruption in recorded history. Over the next few weeks, the volcano would send one hundred cubic kilometers of debris shooting into the air. The plume of dust—1.7 million tons of it—soon spread around the globe, obscuring the sun and causing temperatures to plunge by three to six degrees. There followed several years of severe climate disruption; crops failed around the world, and there were famines in Europe and China; the change in temperature may also have triggered a cholera epidemic in India. In many parts of the world, 1816 would come to be known as the "Year without a Summer."

In May that year, Lord Byron, besieged by scandal, left England and moved to Geneva. He was accompanied by his physician, John Polidori. As it happened, Percy Bysshe Shelley and Mary Wollstonecraft Godwin, who had recently eloped together, were also in Geneva at the time, staying at the same hotel. Accompanying them was Mary Godwin's stepsister, Claire, with whom Byron had had a brief affair in England.

Shelley and Byron met on the afternoon of May 27, and shortly afterward they moved, with their respective parties, to two villas on the shores of Lake Geneva. From there they were able to watch thunderstorms approaching over the mountains. "An almost perpetual rain confines us principally to the house," Mary Shelley wrote. "One night we *enjoyed* a finer storm than I had ever before beheld. The lake was lit up, the pines on the Jura made visible, and all the scene illuminated for an instant, when a pitchy blackness succeeded, and the thunder came in frightful bursts over our heads amid the darkness."

One day, trapped indoors by incessant rain, Byron suggested that they all write ghost stories. A few days later, he outlined an idea for a story "on the subject of the vampyric aristocrat, August Darvell." After eight pages, Byron abandoned the story, and his idea was taken up instead by Polidori: it was eventually published as *The Vampyre* and is now regarded as the first in an ever-fecund stream of fantasy writing.

Mary Shelley too had decided to write a story, and one evening (a stormy one no doubt), the conversation turned to the question of whether "a corpse would be reanimated: galvanism had given token of such things: perhaps the component parts of a creature might be manufactured, brought together and endowed with vital warmth." The next day, she began writing *Frankenstein, or The Modern Prometheus*. Published in 1818, the novel created a sensation: it was reviewed in the best-known journals, by some of the most prominent writers of the time. Sir Walter Scott wrote an enthusiastic review, and he would say later that he preferred it to his own novels. At that time, there does not seem to have been any sense that *Frankenstein* belonged outside the literary mainstream; only later would it come to be regarded as the first great novel of science fiction.

Although Byron never did write a ghost story, he did compose a poem called "Darkness," which was imbued with what we might today call "climate despair":

> The world was void,
> The populous and the powerful—was a lump,
> Seasonless, herbless, treeless, manless, lifeless—
> A lump of death—a chaos of hard clay.
> The rivers, lakes, and ocean all stood still,
> And nothing stirred within their silent depths.

Reflecting on the "wet, ungenial summer" of 1816 and its role in the engendering of these works, Geoffrey Parker writes, "All three works reflect the disorientation and desperation that even a few weeks of abrupt climate change can cause. Since the question today is not *whether* climate change will strike some part of our planet again, but *when*, we might re-read Byron's poem as we choose."

16.

To ask how science fiction came to be demarcated from the literary mainstream is to summon another question: What is it in the nature of modernity that has led to this separation? A possible answer is suggested by Bruno Latour, who argues that one of the originary impulses of modernity is the project of "partitioning," or deepening the imaginary gulf between Nature and Culture: the former comes to be relegated exclusively to the sciences and is regarded as being off-limits to the latter.

Yet, to look back at the evolution of literary culture from this vantage point is to recognize that the project of partitioning has always been contested, and never more so than at the inception, and nowhere more vigorously than in places that were in the vanguard of modernity. As proof of this, we have only to think of William Blake, asking of England:

And was Jerusalem builded here,
Among these dark Satanic mills?

And of Wordsworth's sonnet, "The World Is Too Much With Us":

Little we see in Nature that is ours;
We have given our hearts away, a sordid boon!
. . . .
Great God! I'd rather be
A Pagan suckled in a creed outworn;
So might I, standing on this pleasant lea,
Have glimpses that would make me less forlorn.

Nor was it only in England, but also throughout Europe and North America that partitioning was resisted, under the banners variously of romanticism, pastoralism, transcendentalism, and so on. Poets were always in the forefront of the resistance, in a line that extends from Hölderlin and Rilke to such present-day figures as Gary Snyder and W. S. Merwin.

But being myself a writer of fiction, it is the novel that interests me most, and when we look at the evolution of the form, it becomes evident that its absorption into the project of partitioning was presaged already in the line of Wordsworth's that I quoted above: "I'd rather be / A Pagan suckled in a creed outworn."

It is with these words that the poet, even as he laments the onrushing intrusion of the age, announces his surrender to the most powerful of its tropes: that which envisages time as an irresistible, irreversible forward movement. This jealous deity, the Time-god of modernity, has the power to decide who will be cast into the shadows of backwardness—the dark tunnel of time "outworn"—and who will be granted the benediction of being ahead of the rest, always *en avant*. It is this conception of time (which has much in common with both Protestant and secular teleologies, like those of Hegel and Marx) that allows the work of partitioning to proceed within the novel, always aligning itself with the avant-garde as it hurtles forward in its impatience to erase every archaic reminder of Man's kinship with the nonhuman.

The history of this partitioning is, of course, an epic in itself, offering subplots and characters to suit the tastes of every reader. Here I want to dwell, for a moment, on a plot that completely reverses itself between the eighteenth century and today: the story of the literary tradition's curious relationship with science.

At the birth of modernity, the relationship between literature and science was very close and was perhaps perfectly exemplified in the figure of the writer Bernardin de Saint-Pierre, who wrote one of the earliest of best sellers, *Paul et Virginie*. Saint-Pierre regarded himself as primarily a naturalist and saw no conflict between his calling as writer and man of science. It is said of him that when taken to see the cathedral of Chartres, as a boy, he noticed nothing but the jackdaws that were roosting on the towers.

Goethe also famously saw no conflict between his literary and scientific interests, conducting experiments in optics, and propounding theories that remain compelling to this day. Herman Melville too was deeply interested in the study of marine animals and his views on the subject are, of course, expounded at length in *Moby Dick*. I could cite many other instances ranging from the mathematics of *War and Peace* to the chemistry of *Alice in Wonderland*, but there is no need: it is hardly a matter of dispute that Western writers remained deeply engaged with science through the nineteenth century.

Nor was this a one-sided engagement. Naturalists and scientists not only read but also produced some of the most significant literary works of the nineteenth century, such as Darwin's *Voyage of the Beagle* and Alfred Russell Wallace's *The Malay Archipelago*. Their works, in turn, served as an inspiration to a great number of poets and writers, including Tennyson.

How, then, did the provinces of the imaginative and the scientific come to be so sharply divided from each other? According to Latour the project of partitioning is supported always by a related enterprise: one that he describes as "purification," the purpose of which is to ensure that Nature remains off-limits to Culture, the knowledge of which is consigned entirely to the sciences. This entails the marking off and suppression of hybrids—and that, of course, is exactly the story of the branding of science fiction, as a genre *separate* from the literary mainstream. The line that has been drawn between them exists only for the sake of neatness; because the zeitgeist of late modernity could not tolerate Nature-Culture hybrids.

Nor is this pattern likely to change soon. I think it can be safely predicted that as the waters rise around us, the mansion of serious fiction, like the doomed waterfront properties of Mumbai and Miami Beach, will double down on its current sense of itself, building ever higher barricades to keep the waves at bay.

The expulsion of hybrids from the manor house has long troubled many who were thus relegated to the status of genre writers, and rightly so, for nothing could be more puzzling than the strange conceit that science fiction deals with material that is somehow contaminated; nothing could better express the completeness of the literary mainstream's capitulation to the project of partitioning. And this capitulation has come at a price, for it is literary fiction itself that has been diminished by it. If a list were to be made of the late twentieth-century novelists whose works remain influential today, we would find, I suspect, that many who once bestrode the literary world like colossi are entirely forgotten while writers like Arthur C. Clarke, Raymond Bradbury, and Philip K. Dick are near the top of the list.

That said, the question remains: Is it the case that science fiction is better equipped to address the Anthropocene than mainstream literary fiction? This might appear obvious to many. After all, there is now a new genre of science fiction called "climate fiction" or cli-fi. But cli-fi is made up mostly of disaster stories set in the future, and that, to me, is exactly the rub. The future is but one aspect of the Anthropocene: this era also includes the recent past, and, most significantly, the present.

In a perceptive essay on science fiction and speculative fiction, Margaret Atwood writes of these genres that they "all draw from the same deep well: those imagined other worlds located somewhere apart from our everyday one: in another time, in another dimension, through a doorway into the spirit world, or on the other side of the threshold that divides the known from the unknown. Science Fiction, Speculative Fiction, Sword and Sorcery Fantasy, and Slipstream Fiction: all of them might be placed under the same large 'wonder tale' umbrella."

This lays out with marvelous clarity some of the ways in which the Anthropocene resists science fiction: it is precisely not an imagined "other" world apart from ours; nor is it located in another "time" or another "dimension." By no means are the events of the era of global warming akin to the stuff of wonder tales; yet it is also true that in relation to what we think of as normal now, they are in many ways uncanny; and they have indeed opened a doorway into what we might call a "spirit world"—a universe animated by nonhuman voices.

If I have been at pains to speak of resistances rather than insuperable obstacles, it is because these challenges can be, and have been, overcome in many novels: Liz Jensen's *Rapture* is a fine example of one such; another is Barbara Kingsolver's wonderful novel *Flight Behavior*. Both are set in a time that is recognizable as our own, and they both communicate, with marvelous vividness, the uncanniness and improbability, the magnitude and interconnectedness of the transformations that are now under way.

17.

Global warming's resistance to the arts begins deep underground, in the recesses where organic matter undergoes the transformations that make it possible for us to devour the sun's energy in fossilized forms. Think of the vocabulary that is associated with these substances: *naphtha, bitumen, petroleum, tar,* and *fossil fuels.* No poet or singer could make these syllables fall lightly on the ear. And think of the substances themselves: coal and the sooty residue it leaves on everything it touches; and petroleum—viscous, foul smelling, repellant to all the senses.

Of coal at least it can be said that the manner of its extraction is capable of sustaining stories of class solidarity, courage, and resistance, as in Zola's *Germinal,* for instance, and John Sayles's fine film *Matewan.*

The very materiality of coal is such as to enable and promote resistance to established orders. The processes through which it is mined and transported to the surface create an unusual degree of autonomy for miners; as Timothy Mitchell observes, "the militancy that formed in these workplaces was typically an effort to defend this autonomy." It is no coincidence, then, that coal miners were in the front lines of struggles for the expansion of political rights from the late nineteenth until the mid-twentieth century, and even afterward. It could even be argued that miners, and therefore the economy of coal itself, were largely responsible for the unprecedented expansion of democratic rights that occurred in the West between 1870 and the First World War.

The materiality of oil is very different from that of coal: its extraction does not require large numbers of workers, and since it can be piped over great distances, it does not need a vast workforce for its transportation and distribution. This is probably why its effects, politically speaking, have been the opposite of those of coal. That this might be the case was well understood by Winston Churchill and other leaders of the British and American political elites, which was why they went to great lengths to promote the large-scale use of oil. This effort gained in urgency after the historic strikes of the 1910s and '20s, in which miners, and workers who transported and distributed coal, played a major role; indeed, fear of working-class militancy was one of the reasons why a large part of the Marshall Plan's funds went toward effecting the switch from coal to oil. "The corporatised democracy of postwar Western Europe was to be built," as Mitchell notes, "on this reorganisation of energy flows."

For the arts, oil is inscrutable in a way that coal never was: the energy that petrol generates is easy to aestheticize—as in images and narratives of roads and cars—but the substance itself is not. Its sources are mainly hidden from sight, veiled by technology, and its workers are hard to mythologize, being largely invisible. As for the places where oil is extracted, they possess nothing of the raw visual power that is manifest, for example, in the mining photographs of Sebastião Salgado. Oil refineries are usually so heavily fortified that little can be seen of them other than a distant gleam of metal, with tanks, pipelines, derricks, glowing under jets of flame.

One such fortress figures in my first novel, *The Circle of Reason* (1986), a part of which is about the discovery of oil in a fictional emirate called al-Ghazira: "out of the sand, there suddenly arose the barbed-wire fence of the Oiltown. From the other side of the fence, faces stared silently out—Filipino faces, Indian faces, Egyptian faces, Pakistani faces, even a few Ghaziri faces, a whole world of faces."

Behind these eerie, dislocated enclaves of fenced-in faces and towering derricks lies a history that impinges on every life on this planet. This is true especially in regard to the Arabian peninsula, where oil brought about an encounter with the West that has had consequences that touch upon every aspect of our existence, extending from matters of security to the buildings that surround us and the quality of the air we breathe. Yet the strange reality is that this historic encounter, whose tremors and aftershocks we feel every day, has almost no presence in our imaginative lives, in art, music, dance, or literature.

Long after the publication of *The Circle of Reason*, I wrote a piece in which I attempted to account for this mysterious absence: "To the principal protagonists in the Oil Encounter (which means, in effect, America and Americans, on the one hand, and the peoples of the Arabian peninsula and the Persian Gulf, on the other), the history of oil is a matter of embarrassment verging on the unspeakable, the pornographic. It is perhaps the one cultural issue on which the two sides are in complete agreement. . . . Try and imagine a major American writer taking on the Oil Encounter. The idea is literally inconceivable."

The above passage figures in a review of one of the few works of fiction to address the Oil Encounter, a five-part series of novels by the Jordanian-born writer Abdel Rahman Munif. My review, entitled "Petrofiction," dealt only with the first two books in the cycle, which were published in English translation as *Cities of Salt* (*Mudun al Malh*) and *The Trench* (*Al-ukhdud*).

"The truth is," I wrote in my review, "that we do not yet possess the form that can give the Oil Encounter a literary expression. For this reason alone *Cities of Salt* . . . ought to be regarded as a work of immense significance. It so happens that the first novel in the cycle is also in many ways a wonderful work of fiction, perhaps even in parts a great one."

This was written in 1992. I was not aware then that *Cities of Salt* had been reviewed four years earlier by one of the most influential figures in the American literary firmament: John Updike. His review, when I read it, made a great impression on me: I found that in the process of writing about Munif's book Updike had also articulated, elegantly and authoritatively, a conception of the novel that was indisputably an accurate summing-up of a great deal of contemporary fiction. Yet it was a conception with which I found myself completely at odds.

The differences between Updike's views and mine have an important bearing on some of the aspects of the Anthropocene that I have been addressing here, so it is best to let him speak for himself. Here is what he had to say about *Cities of Salt*: "It is unfortunate, given the epic potential of his topic, that Mr. Munif . . . appears to be . . . insufficiently Westernized to produce a narrative that feels much like what we call a novel. His voice is that of a campfire explainer; his characters are rarely fixed in our minds by a face or a manner or a developed motivation; no central figure develops enough reality to attract our sympathetic interest; and, this being the first third of a trilogy, what intelligible conflicts and possibilities do emerge remain serenely unresolved. There is almost none of that sense of individual moral adventure—of the evolving individual in varied and roughly equal battle with a world of circumstance—which since 'Don Quixote' and 'Robinson Crusoe,' has distinguished the novel from the fable and the chronicle; 'Cities of Salt' is concerned, instead, with men in the aggregate."

This passage is remarkable, in the first instance, because the conception of the novel that is articulated here is rarely put into words, even though it has come to exercise great sway across much of the world and especially in the Anglosphere. My own disagreement with it hinges upon the phrase that Updike uses to distinguish the novel from the fable and the chronicle: "individual moral adventure."

But why, I find myself asking, should the defining adventures of the novel be described as "moral," as opposed to, say, intellectual or political or spiritual? In what sense could it be said that *War and Peace* is about individual moral adventures? It is certainly true that some threads in the narrative could be described in this way, but they would account for only a small part of the whole. As for Tolstoy's own vision of what he had set out to do, he was emphatic that *War and Peace* was "not a novel, still less a long poem, and even less a historical chronicle": his intention was to supersede and incorporate preceding forms— an ambition that can be seen also in Melville's *Moby Dick*. To fit these works within the frame of "individual moral adventures" is surely a diminishment of the writers' intentions.

Clearly, it is in the word *moral* that the conundrum lies: What exactly does it mean? Is it intended perhaps to incorporate the senses also of the "political," the "spiritual," and the "philosophical"? And if so, would not a question arise as to whether a single word can bear so great a burden?

I ask these questions not in order to parse small semantic

differences. I believe that Updike had actually put his finger on a very important aspect of contemporary culture. I will return to this later, but for now I'd like to turn to another aspect of Updike's mapping of the territory of the novel: that which is excluded from it.

Updike draws this boundary line with great clarity: the reason why *Cities of Salt* does not feel "much like a novel," he tells us, is that it is concerned not with a sense of individual moral adventures but rather with "men in the aggregate." In other words, what is banished from the territory of the novel is precisely the collective.

But is it actually the case that novelists have shunned men (or women) in the aggregate? And inasmuch as they have, is it a matter of intention or narrative expediency? Charlotte Brontë's view, expressed in a letter to a critic, is worth noting: "is not the real experience of each individual very limited?" she asks, "and if a writer dwells upon that solely or principally is he not in danger of being an egotist?"

In a perceptive discussion of Updike's review, the critic Rob Nixon points out that Munif is "scarcely alone in working with a crowded canvas and with themes of collective transformation"; Émile Zola, Upton Sinclair, and many others have also treated "individual character as secondary to collective metamorphosis."

Indeed, so numerous are the traces of the collective within the novelistic tradition that anyone who chose to look for them would soon be overwhelmed. Such being the case, should Updike's view be summarily dismissed? My answer is: no—because Updike was, in a certain sense, right. It is a fact that the contemporary novel has become ever more radically centered on the individual psyche while the collective—"men in the aggregate"—has receded, both in the cultural and the fictional imagination. Where I differ from Updike is that I do not think that this turn in contemporary fiction has anything to do with the novel as a form: it is a matter of record that historically many novelists from Tolstoy and Dickens to Steinbeck and Chinua Achebe have written very effectively about "men in the aggregate." In many parts of the world, they continue to do so even now.

What Updike captures, then, is not by any means an essential element of the novel as a form; his characterization is true rather of a turn that fiction took at a certain time in the countries that were then leading the way to the "Great Acceleration" of the late twentieth century. It is certainly no coincidence that these were the very places where, as Guy Debord observed, the reigning economic system was not only founded on isolation, it was also "designed to produce isolation."

I say it is no coincidence for two reasons. The first is that the acceleration in carbon emissions and the turn away from the collective are both, in one sense, effects of that aspect of modernity that sees time (in Bruno Latour's words) as "an irreversible arrow, as capitalization, as progress." I've noted before that this idea of a continuous and irreversible forward movement, led by an avant-garde, has been one of the animating forces of the literary and artistic imagination since the start of the twentieth century. A progression of this sort inevitably creates winners and losers, and in the case of twentieth-century fiction, one of the losers was exactly writing of the kind in which the collective had a powerful presence. Fiction of this sort was usually of a realist variety, and it receded because it was consigned to the netherworld of "backwardness."

But the era of global warming has made audible a new, nonhuman critical voice that forces us to ask whether those old realists were so "used-up" after all. Consider the example of John Steinbeck, never a favorite of the avant-garde, and once famously dismissed by Lionel Trilling as a writer who thought "like a social function, not a novelist." Yet, if we look back upon Steinbeck now, in full awareness of what is now known about the future of the planet, his work seems far from superseded; quite the contrary. What we see, rather, is a visionary placement of the human within the nonhuman; we see a form, an approach that grapples with climate change avant la lettre.

Around the world too there are many writers—not all of them realists—from whose work neither the aggregate nor the nonhuman have ever been absent. To cite only a few examples from India: in Bengali, there is the work of Adwaita Mallabarman and Mahasweta Devi; in Kannada, Sivarama Karanth; in Oriya, Gopinath Mohanty; in Marathi, Vishwas Patil. Of these writers too I suspect that Updike would have said that their books were not much "like what we call a novel."

But once again, the last laugh goes to that sly critic, the Anthropocene, which has muddied, and perhaps even reversed, our understanding of what it means to be "advanced." Were we to adopt the arrow-like time perspective of the moderns, there is a sense in which we might even say of writers like Munif and Karanth that they were actually "ahead" of their peers elsewhere.

Here, then, is another reason why something more than mere chance appears to be at play in the turn that fiction took as emissions were rising in the late twentieth century. It is one of the many turns of that period that give, in retrospect, the uncanny impression that global warming has long been toying with humanity (thus, for example, the three postwar decades, when emissions grew sharply, saw a *stabilization* of global temperatures). Similarly, at exactly the time when it has become clear that global warming is in every sense a collective predicament, humanity finds itself in the thrall of a dominant culture in which the idea of the collective has been exiled from politics, economics, and literature alike.

Inasmuch as contemporary fiction is caught in this thrall-
dom, this is one of the most powerful ways in which global
warming resists it: it is as if the gas had run out on a generation accustomed to jet skis, leaving them with the task of re-
inventing sails and oars.

18.

Mrauk-U is the site of a vast and enchanting complex of Buddhist pagodas and monasteries in western Burma. Once the
capital of the Arakan (Rakhine) kingdom, it flourished between the fourteenth and seventeenth centuries. During that
time, it was an important link in the networks of trade that
spanned the Indian Ocean. Chinese ceramics and Indian tex-
tiles passed through it in quantity; merchants and travelers
came from Gujarat, Bengal, East Africa, Yemen, Portugal, and
China to sojourn in the city. The wealth they brought in allowed the kingdom's rulers to honor their religion by embark-
ing on vast building projects. The site they created is smaller
than Bagan or Angkor Wat, but it is, in its own way, just as
interesting.

Getting to Mrauk-U isn't easy. The nearest town of any size
is Sittwe (formerly Akyab), and from there the journey to the
site can take a day or more, depending on the condition of the
road. As Mrauk-U approaches, ranges of low hills, of rounded,
dome-like shapes appear in the distance; at times, the ridg-
es seem to rise into spires and finials. Such is the effect that
the experience of entering the site is like stepping into a zone
where the human and nonhuman echo each other with an
uncanny resonance; the connection between built form and
landscape seems to belong to a dimension other than the visual; it is like that of sympathetic chords in music. The echoes
reach into the interiors of the monuments, which, with their

openings and pathways, their intricate dappling of light and shadow, and their endless iterations of images, seem to aspire to be forests of stone.

In *How Forests Think*, the anthropologist Eduardo Kohn suggests that "forms"—by which he means much more than shapes or visual metaphors—are one of the means that enable our surroundings to think through us.

But how, we might ask, can any question of thought arise in the absence of language? Kohn's answer is that to imagine these possibilities we need to move beyond language. But to what? Merely to ask that question is to become aware of the multiple ways in which we are constantly engaged in patterns of communication that are not linguistic: as, for example, when we try to interpret the nuances of a dog's bark; or when we listen to patterns of birdcalls; or when we try to figure out what exactly is portended by a sudden change in the sound of the wind as it blows through trees. None of this is any less demanding, or any less informative, than, say, listening to the news on the radio. We do these things all the time—we could not stop doing them if we tried—yet we don't think of them as communicative acts. Why? Is it perhaps because the shadow of language interposes itself, preventing us from doing so?

It isn't only the testimony of our ears that is blocked in this way, but also that of our eyes, for we often communicate with animals by means of gestures that require interpretation—as, for example, when I wave my hands to shoo away a crow. Nor does interpretation necessarily demand a sense of hearing or sight. In my garden, there is a vigorously growing vine that regularly attempts to attach itself to a tree, several meters away, by "reaching" out to it with a tendril. This is not done randomly, for the tendrils are always well aimed and they appear at exactly those points where the vine does actually stand a chance of bridging the gap: if this were a human, we would say that she was taking her best shot. This suggests to me that the vine is, in a sense, "interpreting" the stimuli around it, perhaps the shadows that pass over it or the flow of air in its surroundings. Whatever those stimuli might be, the vine's "reading" of them is clearly accurate enough to allow it to develop an "image" of what it is "reaching" for; something not unlike "heat-imaging" in weapons and robots.

To think like a forest, then, is, as Kohn says, to think in images. And the astonishing profusion of images in Mrauk-U, most of which are of the Buddha in the *bhumisparshamudra*, with the tip of the middle finger of his right hand resting on the earth, serves precisely to direct the viewer away from language toward all that cannot be "thought" through words.

These possibilities have, of course, been explored by people in many cultures and in many eras—in fact, everywhere perhaps except within the modern academy. What, then, is to be made of the fact that such possibilities have now succeeded also in broaching the boundaries of the one sphere from which they were excluded? Could it be said, extending Kohn's argument, that this synchronicity confirms that the Anthropocene has become our interlocutor, that it is indeed thinking "through" us? Would it follow, then, on the analogy of Kohn's suggestion in relation to forests, that to think about the Anthropocene will be to think in images, that it will require a departure from our accustomed logocentricism? Could that be the reason why television, film, and the visual arts have found it much easier to address climate change than has literary fiction? And if that is so, then what does it imply for the future of the novel?

It is possible, of course, to construct many different genealogies for the deepening logocentricism of the last several centuries. But the one point where all those lines of descent converge is the invention of print technology, which moved the logocentricism of the Abrahamic religions in general, and the Protestant Reformation in particular, onto a new plane. So much so that Ernest Gellner was able to announce in 1964, "The humanist intellectual is, essentially, an expert on the written word."

Merely to trace the evolution of the printed book is to observe the slow but inexorable excision of all the pictorial elements that had previously existed within texts: illuminated borders, portraits, coloring, line drawings, and so on. This pattern is epitomized by the career of the novel, which in the eighteenth and nineteenth centuries often included frontispieces, plates, and so on. But all of these elements gradually faded away, over the course of the nineteenth and early twentieth centuries, until the very word *illustration* became a pejorative, not just within fiction but in all the arts. It was as if every doorway and window that might allow us to escape the confines of language had to be slammed shut, to make sure that humans had no company in their dwindling world but their own abstractions and concepts. This, indeed, is a horizon within which every advance is achieved at the cost of "making the world more unlivable."

But then came a sea change: with the Internet we were suddenly back in a time when text and image could be twinned with as much facility as in an illuminated manuscript. It is surely no coincidence that images too began to seep back into the textual world of the novel; then came the rise of the graphic novel—and it soon began to be taken seriously.

So if it is the case that the last, but perhaps most intransigent way the Anthropocene resists literary fiction lies ultimately in its resistance to language itself, then it would seem to follow that new, hybrid forms will emerge and the act of reading itself will change once again, as it has many times before.

PART II

History

1.

In accounts of the Anthropocene, and of the present climate crisis, capitalism is very often the pivot on which the narrative turns. I have no quarrel with this: as I see it, Naomi Klein and others are right to identify capitalism as one of the principal drivers of climate change. However, I believe that this narrative often overlooks an aspect of the Anthropocene that is of equal importance: empire and imperialism. While capitalism and empire are certainly dual aspects of a single reality, the relationship between them is not, and has never been, a simple one: in relation to global warming, I think it is demonstrably the case that the imperatives of capital and empire have often pushed in different directions, sometimes producing counter-intuitive results.

To look at the climate crisis through the prism of empire is to recognize, first, that the continent of Asia is conceptually critical to every aspect of global warming: its causes, its philosophical and historical implications, and the possibility of a global response to it. It takes only a moment's thought for this to be obvious. Yet, strangely, the implications are rarely reckoned with—and this may be because the discourse around the Anthropocene, and climate matters generally, remains largely Eurocentric. This is why the case for Asia's centrality to the climate crisis does need to be laid out in some detail, even if it is at the cost of stating the obvious.

2.

Asia's centrality to global warming rests, in the first instance, upon numbers. The significance of this is perhaps most readily apparent in relation to the future; that is to say, if we consider the location of those who are most at threat from the changes that are now under way across the planet. The great majority of potential victims are in Asia.

The effect of mainland Asia's numbers is such as to vastly amplify the human impacts of global warming. Take, for instance, the Bengal Delta (a region that consists of most of Bangladesh and much of the Indian state of West Bengal). Formed by the confluence of two of the world's mightiest rivers, the Ganges and the Brahmaputra, this is one of the most densely populated parts of the world, with more than 250 million people living in an area about a quarter the size of Nigeria.

The floodplains of Bengal are not likely to be submerged as soon or as completely as, say, the Pacific island nation of Tuvalu. But the population of Tuvalu is less than ten thousand while the partial inundation of just one island in Bangladesh— Bhola island—has led to the displacement of more than half a million people.

Because of the density of its population, some of the world's worst disasters have occurred in the Bengal Delta. The 1971 Bhola cyclone is thought to have killed three hundred thousand people. As recently as 1991, a cyclone in Bangladesh resulted in 138,000 dead, of whom 90 percent were women. Sea-level rise and the increasing intensity of storms will make large-scale inundations more likely, all along the coastline.

Moreover, in Bengal, as in other Asian deltas, for example, those of the Irrawaddy, the Indus, and the Mekong, another factor has magnified the effects of sea-level rise: this is that delta regions across Asia (and elsewhere in the world) are subsiding much faster than the oceans are rising. This is due partly to geological processes and partly to human activities, such as dam building and the extraction of groundwater and oil. Again, the southern parts of Asia are particularly vulnerable, with the deltas of the Chao Phraya, the Krishna-Godavari, the

Ganges-Brahmaputra, and the Indus being especially imperiled. The Indus, on which Pakistan is critically dependent, has been exploited to the point where it no longer reaches the sea and, as a result, salt water has pushed inland by forty miles, swallowing up over a million acres of agricultural land.

In India a significant rise in sea level could lead to the loss of some six thousand square kilometers, including some of the country's most fertile lands; many of the subcontinent's low-lying islands, like the Lakshadweep chain, may disappear. One study suggests that rising sea levels could result in the migration of up to 50 million people in India and 75 million in Bangladesh. Along with Bangladesh, Vietnam is at the top of the list of countries threatened by sea-level rise: in the event of a one-meter rise in sea level, more than a tenth of Vietnam's population will be displaced.

The ongoing changes in climate pose a dire threat also to the interior of the continent where millions of lives and livelihoods are already in jeopardy because of droughts, periodic flooding, and extreme weather events. No less than 24 percent of India's arable land is slowly turning into desert, and a two-degree Celsius rise in global average temperature would reduce the country's food supply by a quarter. In Pakistan, a hundred thousand acres of salt-encrusted land are being abandoned each year; of the fields that remain "a fifth are badly waterlogged and a quarter produce only meagre crops." In China, which feeds more than 20 percent of the world's population off 7 percent of the world's arable land, desertification is already causing direct annual losses of $65 billion.

Fearsome as these risks are, they are dwarfed by Asia's accelerating water crisis. The rivers that sustain China and South and Southeast Asia rise in Tibet and the Himalayas; the waters that are stored there, in the form of accumulations of ice, sustain 47 percent of the world's population: "here the water-related dreams and fears of half the human race come together." But this region is warming twice as fast as the average global rate, and in 2008 it was found that the Himalayan glaciers had already lost *all* the ice formed since the mid-1940s; by some reckonings, one-third of them will disappear by 2050.

As the melting of the Himalayan glaciers accelerates, the variations in the rivers' flow will increase, falling to unprecedented lows in the dry season and causing massive inundations in the summer, as in the Kosi River disaster of 2008 in Bihar, and the Indus floods of 2010. And if the glaciers continue to shrink at the present rate, the most populous parts of Asia will face catastrophic water shortages within a decade or two. A quarter of the world's rivers already run dry before reaching the sea: many, if not most, of them are in Asia.

In terms of numbers, the consequences are beyond imagining: the lives and livelihoods of half a billion people in South and Southeast Asia are at risk. Needless to add, the burden of these impacts will be borne largely by the region's poorest people, and among them disproportionately by women.

It is the matter of numbers again that makes Asia critical to the questions of mitigation, preparedness, and resiliency. Aquifers are drying up in northern China as well as in America's Great Plains: but only 2 million people live in the 175,000 square miles that are watered by the United States' Ogallala Aquifer while the 125,000 square miles of north China are populated by 214 million people.

The brute fact is that no strategy can work globally unless it works in Asia and is adopted by large numbers of Asians. Yet, in this matter too, the conditions that are peculiar to mainland Asia are often absent from the discussion.

3.

The vulnerability of Asia's populations is only one aspect of their centrality to global warming. The reality is that the continent has also played a pivotal role in setting in motion the chain of consequences that is driving the present cycle of climatic change. In this story, too, numbers are critical, for it was the rapid and expanding industrialization of Asia's most populous nations, beginning in the 1980s, that brought the climate crisis to a head.

Numbers are critical again to the difference in Asia's role in global warming and that of countries that industrialized earlier. The West's largest contribution to the accumulation of greenhouse gases came about through the continuous expansion of the carbon footprint of what was about 30 percent of the world's population at the beginning of the twentieth century. Asia's contribution, on the other hand, came about through a sudden but very small expansion in the footprint of a much larger number of people, perhaps as much as half of a greatly expanded global population, late in the twentieth century.

To be sure, the planet would have faced a climate crisis sooner or later, even if the history of mainland Asia had not taken this turn. After all, signs of a changing climate date back to the 1930s, and the concentration of carbon dioxide in the atmosphere had already passed 300 parts per million when Charles Keeling began to take measurements at the Mauna Loa Observatory in Hawaii. This was in the late 1950s, long before the economies of mainland Asia began their rapid acceleration. Even back then, the carbon footprint of the West was growing rapidly enough to ensure that the accumulation of greenhouse gases in the atmosphere would continue to rise. But that rise would not have been so steep if mainland Asia had not launched upon a period of sustained economic expansion in the late 1980s. It is this acceleration that has dramatically shortened the time available to adapt to, or even recognize, the crisis for what it is.

But apart from its dual role as both protagonist and victim, Asia has played yet another critical part in the unfolding of the Great Derangement: it is that of the simpleton who, in his blundering progress across the stage, unwittingly stumbles upon the secret that is the key to the plot. This is because certain crucial aspects of modernity would not have become apparent if they had not been put to an empirical test, in the only continent where the magnitudes of population are such that they can literally move the planet. And as with any truly revelatory experiment, the results would not have been believed, by Asians or anyone else, if they had not turned out exactly as they have. For the results are counterintuitive and they contradict all the tenets on which our lives, thoughts, and actions have been based for almost a century. What we have learned from this experiment is that the patterns of life that modernity engenders can only be practiced by a small minority of the world's population. Asia's historical experience demonstrates that our planet will not allow these patterns of living to be adopted by every human being. Every family in the world cannot have two cars, a washing machine, and a refrigerator—not because of technical or economic limitations but because humanity would asphyxiate in the process.

It is Asia, then, that has torn the mask from the phantom that lured it onto the stage of the Great Derangement, but only to recoil in horror at its own handiwork; its shock is such that it dare not even name what it has beheld—for having entered this stage, it is trapped, like everyone else. All it can say to the chorus that is waiting to receive it is "But you promised . . . and we believed you!"

In this role as horror-struck simpleton, Asia has also laid bare, through its own silence, the silences that are now ever more plainly evident at the heart of global systems of governance.

4.

If it is the case that the climate crisis was precipitated by mainland Asia's embrace of the dominant mechanisms of the world economy, then the critical question in relation to the history of the Anthropocene is this: Why did the most populous countries of Asia industrialize late in the twentieth century and not before?

Strangely this question is almost never explicitly posed in accounts of the history of global warming. Yet these histories do often offer an implicit answer to the question of why the non-Western world was slow to enter the carbon economy: it is simply that the technologies that created this economy (e.g., the spinning jenny and the steam engine) were invented in England and were therefore inaccessible to much of the world. In this view industrialization comes about through a process of technological diffusion that radiates outward from the West.

This narrative is, of course, consistent with the history of global warming over the nineteenth and twentieth centuries, when the carbon-intensive economies of the West pumped greenhouse gases into the atmosphere at ever accelerating rates. It is therefore perfectly accurate to say, as Anil Agarwal and Sunita Narain did in their seminal 1991 essay on climate justice, that "the accumulation in the atmosphere of [greenhouse] gases is mainly the result of the gargantuan consumption of the developed countries, particularly the U.S." Yet these truths should not lead us to overlook the fact that this economy had a very complicated prehistory.

Before the advent of the carbon-intensive economy, the populations of the "old world" were not divided by vast gaps in technology. For millennia, trade connections were close enough to ensure that innovations in thought and technique were transmitted quite rapidly over long distances. Even "deep," long-term historical processes sometimes unfolded at roughly the same time in places far removed from each other. The vernacularization of languages is an example of one such: as Sheldon Pollock has shown, this process began almost simultaneously in Europe and the Indian subcontinent. The stimulus may also have been the same in both instances, consisting of forces set in motion by the Islamic expansion.

There is now a great deal of research to suggest that the early modern period, roughly the sixteenth to the early nineteenth centuries, was a time of rapid and often parallel change around much of the world, and particularly so across the Eurasian landmass. The fact that these developments were set in motion during a period of great climatic disruption (i.e., the seventeenth century) opens the door to the possibility that the changes of the early modern era were influenced by shifts in climate, which had varying effects on different parts of the planet.

It is a fact, in any event, that exchanges of technology and knowledge accelerated in the early modern period. In the sixteenth century, for instance, innovations in weaponry and fortifications traveled very quickly between Europe, the Middle East, and India. The same was true of ideas: early botanical works, like the seventeenth-century *Hortus Malabaricus*, were often produced in collaboration by Europeans and savants from elsewhere. A continuous cross-pollination of ideas occurred in mathematics too. It is now known that the Kerala School of Mathematics anticipated the work of "Gregory, Newton and Leibniz by at least 250 years"; it is by no means unlikely that these developments were transmitted to Europe by Jesuits. Although non-Western influences usually went unacknowledged in Europe, in at least one instance, that of the nineteenth-century logician and mathematician George Boole, they *were* explicitly recognized. Boole's wife, Mary Everest Boole, even made the claim that nineteenth-century European science "could never have reached its present height had it not been fertilized by successive wafts from the . . . knowledge stored up in the East."

Philosophy provides a particularly interesting example, both of parallel development and of the circulation of ideas. As the philosopher Jonardon Ganeri has shown, the innovations of the Nava-Nyaya school of philosophy in Bengal contain striking similarities to the thought of early modern philosophers in Europe. So rapid was the circulation of philosophical ideas that Muslim, Jain, and Hindu philosophers were familiar with the ideas of Descartes within "ten years after his death." A major figure in the transmission and circulation of these ideas was the French traveler François Bernier, who translated Descartes into Persian during his travels in Asia. So great was the ferment of this period, writes Ganeri, that "India in the seventeenth century . . . was in intellectual overdrive. Muslim, Jaina, and Hindu intellectuals produced work of tremendous vitality, and ideas circulated around India, through the Persianate and Arabic worlds, and out to Europe and back."

In short, as the historian Sanjay Subrahmanyam has long argued, modernity was not a "virus" that spread from the West to the rest of the world. It was rather a "global and conjunctural phenomenon," with many iterations arising almost simultaneously in different parts of the world.

That such a possibility might exist had long been obscured by one of the distinctive features of Western modernity: its insistence on its own uniqueness. Yet even this insistence has usually been suspended in relation to one case: that of Japan, of which it is widely accepted that it had its own unique variant of modernity. This was due, no doubt (as Subrahmanyam suggests), to Japan's bank balance, which was large enough to legitimize its claims to a singular form of modernity. But now, with the swelling of the bank balances of India, China, and many other nations, it is increasingly apparent that the early modern era nurtured not one or two but "multiple modernities."

This multiplicity extended also to the use of fossil fuels, which has a long non-Western history. Although largely forgotten now, this history provides some revealing insights into the emergent modernities of the period leading up to and immediately after the Industrial Revolution.

5.

About a thousand years ago, China went through a "medieval economic revolution." This led to so much deforestation that the resulting run-off of silt actually altered the coastline, leading to the filling-in and expansion of the deltas of the Pearl, Yellow, and Yangtze Rivers. By the eleventh century, the shortage of wood was such that the people of northern Jiangsu province were overjoyed to learn of the discovery of coal in their region. This prompted the poet Su Dongpo to write the following lines in 1087 CE:

> That a rich inheritance lay in their hills was something they did not know:
> Lovely black rock in abundance, ten thousand cartloads of coal

No one had noticed the spatters of tar, nor the bitumen,
 where it oozed leaking,
While puff after puff, the strong-smelling vapors—drifted
 off on their own with the breezes

Once the leads to the seams were unearthed, it was found
 to be huge and unlimited
People danced in throngs in their jubilation. Large numbers
 went off to visit it.

In short, coal has been used and appreciated in China for a
very long time. Why then did China not make the transition
to a large-scale coal economy before Britain? The work of the
historian Kenneth Pomeranz suggests that the answer may
lie in a purely contingent factor: it may simply be that China's
coal reserves, unlike Britain's, were not in easily accessible locations.

But the Chinese were also pioneers in the use of other fossil
fuels. This is what the late eighteenth-century manual *Classic
of the Waterways of Sichuan* has to say about the use and extraction of oil and natural gas in that province:

[Natural gas] can be used as fuel to boil brine, steam rice,
calcine limestone to make lime, or to smother-burn wood to
convert it to charcoal. When it is drawn through an opening in a bamboo tube to serve as a substitute for firewood
or candles, this aperture is smeared with clay. Thus it burns
hotly at the mouth but the bamboo does not catch fire. At
other times it is drawn off into the bladder of a pig, the
opening sealed, and the bladder placed inside a box or bag
that one can take home with one. When night falls, one
pierces a hole with a needle, applies an ordinary domestic

flame, and fire will come out of the bladder and light up the room.

There are also oil wells. The color of the oil is turbid, but it burns well and one can have a fire anywhere. It will not diminish or go out in wind or rain, or even when plunged into water. If one is traveling at night, and stores oil in a bamboo tube, it is possible to travel for one or two kilometers on a single tube. . . . Such wells are commonplace, and no cause for astonishment.

Mark Elvin (from whose book, *The Retreat of the Elephants*, these excerpts are taken) further notes, "Parts of this region of late-premodern China not only used gas for industrial processes, but had domestic gas cookers, domestic gas lighting, and a primitive form of mobile illumination using bottled oil. All based on bamboo tubing. . . . Once again . . . there is something of the feel of an emerging modern economy about this technical vigor and virtuosity."

6.

My novel *The Glass Palace* touches upon a place in Burma where oil had for many centuries bubbled up to the surface and formed rivulets. The place is called Yenangyaung, and it takes its name from the foul smell of the ooze.

Of all the river's sights the strangest was one that lay a little to the south of the great volcanic hump of Mount Popa. The Irrawaddy here described a wide, sweeping turn, spreading itself to a width of over a mile. On the eastern bank of the river, there appeared a range of low, foul-smelling mounds. These hillocks were covered in a thick ooze, a substance that would sometimes ignite spontaneously in the heat of the sun, lighting fires that trickled slowly to the water's edge. Often at night small, wavering flames could be seen in the distance, carpeting the slopes.

To the people of the area this ooze was known as earth-oil: it was a dark, shimmering green, the colour of bluebottles' wings. It seeped from the rocks like sweat, gathering in shiny green-filmed pools. In places, the puddles joined together to form creeks and rivulets, an oleaginous delta that fanned out along the shores. So strong was the odour of this oil that it carried all the way across the Irrawaddy: boatsmen would swing wide when they floated past these slopes, this place-of-stinking-creeks—Yenangyaung.

This was one of the few places in the world where petroleum rose naturally to the surface of the earth. Long before the discovery of the internal combustion engine there was already a good market for this oil. . . . Merchants came to Yenangyaung from as far away as China to avail themselves of this substance. The gathering of the oil was the work of a community endemic to those burning hills, a group of people known as *twin-zas*, a tight-knit, secretive bunch of outcasts, runaways and foreigners.

Over generations *twin-za* families had attached themselves to individual springs and pools, gathering the oil in buckets and basins, and ferrying it to nearby towns. Many of Yenangyaung's pools had been worked for so long that the level of oil had sunk beneath the surface, forcing their owners to dig down. In this way, some of the pools had gradually become wells, a hundred feet deep or even more—great oil-sodden pits, surrounded by excavated sand and earth. Some of these wells were so heavily worked that they looked like small volcanoes, with steep, conical slopes. At these depths the oil could no longer be collected simply by dipping a weighted bucket: *twin-zas* were lowered in, on ropes, holding their breath like pearl-divers.

. . . The rope would be attached, by way of a pulley, to [the *twin-za's*] wife, family and livestock: they would lower him in by walking up the slope of the well, and when they felt his tug they would pull him out again by walking down. The lips of the wells were slippery from spills and it was not uncommon for unwary workers and young children to tumble in. Often these falls went unnoticed: there were no splashes and few ripples. Serenity is one of the properties of this oil: it is not easy to make a mark upon its surface.

These paragraphs describe Yenangyaung as it was in the latter years of nineteenth century. But the history of Burma's oil industry goes back much further, possibly even a millennium or more.

Oil from natural springs, sinks, and hand-dug pits has of course been used in many parts of the world since ancient times. But it is quite likely that "the early oil industry of Burma" was "the largest in the world."

The oil wells of Yenangyaung caught the attention of British travelers as early as the mid-eighteenth century. Major Michael Symes, an East India Company envoy to the court of Ava, published this description of it in 1795:

After passing various sands and villages, we got to Yaynangheoum or Earthoil (Petroleum) Creek about two hours past noon. . . . We were informed, that the celebrated wells of Petroleum, which supply the empire [of Ava], and many parts of India, with that useful product, were five miles to the east of this place. . . . The mouth of the creek was crowded with large boats, waiting to receive a lading of oil; and immense pyramids of earthen jars were raised within and round the village, disposed in the same manner as shot and shells are piled in an arsenal. . . . We saw several thousand jars filled with it ranged along the bank.

The earth-oil of Yenangyaung had many uses: it was applied on the skin as a remedy for certain conditions, and was also used as an insecticide, as a lubricant for cartwheels, as a caulking-agent in boatbuilding, and even as a preservative for palm-leaf manuscripts. But its principal use was as a fuel for lamps; in 1826, a British official was told that two-thirds of the oil of Yenangyaung served that purpose. It was this form of illumination that sustained the nighttime *pwes* or festivals that were, and still are, so beloved by the Burmese.

For the ruling Konbaung dynasty, the oil industry was a major source of revenue even in the eighteenth century, which was one reason why British envoys took a special interest in it. But oil became especially important after the Second Anglo-Burmese War of 1852–53, when the British seized a large part of the kingdom, depriving the then ruler, King Mindon, of his southern revenues. This greatly increased the king's dependence on oil, and in 1854 he did what the rulers of many modern petro-states were to do in the century to come: he asserted direct control over the oil fields of Yenangyaung, effectively nationalizing the industry. After this, producers could sell only to the state, and the king was able to dictate prices. At the same time, King Mindon also took steps to create links with the world market by entering into contracts with an English firm that manufactured paraffin candles. Soon, Price's Patent Candle Company Ltd. was importing approximately two thousand barrels of Burmese oil per month. This amounted to more than half of the yearly production of the Yenangyaung oil fields, which was in the range of forty-six thousand barrels.

The king further consolidated his relationship with the oil

industry by marrying the daughter of a leading well-owner, thereby acquiring control of over 120 oil wells. King Mindon is also said to have created a refinery in Mandalay, where oil was stored and processed. These and other interventions brought about a twofold increase in oil production in Burma between 1862 and 1876.

In light of this, it could be said that the first steps toward the creation of a modern oil industry were actually taken in Burma. But where these steps might have led we do not know because Burma's attempts to control its oil came to an abrupt end in 1885, when the British invaded and annexed the remnants of the Konbaung realms, deposing Thibaw, the dynasty's · last king. After that, the oil fields of Yenangyaung passed into British control, and, in time, they became the nucleus of the megacorporation that was known until the 1960s as Burmah-Shell. But throughout the nineteenth and twentieth centuries, the *twin-za*s of Yenangyaung continued to play a major role in exploiting the oil fields of the region: they do so to this day.

In nineteenth-century British sources, the unfortunate Konbaungs are often represented as indolent, corrupt, and backward. But as the historian Thant Myint-U has shown, the late Konbaungs—even the doomed Thibaw—were, in their own fashion, trying hard to stay abreast in matters of technology: they introduced the telegraph, imported steam-powered vessels, enacted administrative reforms of many kinds, encouraged manufacturing industries, tried to build railways, and created scholarships to educate students in France and England. They also exerted themselves to improve animal welfare, in line with Buddhist teachings. At King Mindon's behest, a number of wildlife sanctuaries were established in the Burmese kingdom from the 1850s onward.

There is no reason to suppose that Burma would have been unable to navigate the emerging petroleum economy had it been free to do so: certainly in the mid-nineteenth century no part of the world had more experience in the production of oil than Burma.

It will be clear from this that, as with much else that bears the label of "modern," the development of the oil industry in Burma was a profoundly hybrid process, involving local rulers, officialdom, and businessmen, not to speak of technologies that dated back many centuries. Yet, as the historian Marilyn Longmuir notes, "Most oil historians give the date 28 August 1859 as the commencement of the modern oil industry when 'Colonel' Edwin L. Drake organized the first successful drilling of an oil well at Oil Creek near Titusville, Pennsylvania."

Here again is an instance of what I cited earlier as the one feature of Western modernity that is truly distinctive: its enormous intellectual commitment to the promotion of its supposed singularity.

7.

The first steam-powered vessel to operate in India was a dredger on the Hooghly River. The vessel's engine is said to have been sent from Birmingham to Calcutta in 1817 or 1818, little more than a decade after Robert Fulton made history by launching the first commercial steamboat on the Hudson River in 1807.

The first marine steam engines to see commercial service in India were purchased by a group of Calcutta businessmen from a British trader in Canton in 1823. The two engines were mounted on a locally built vessel, which was launched upon the Hooghly under the name *Diana*. Although the *Diana* attracted much attention, the venture was a commercial failure. But the import of steam engines continued at a steady pace:

the records of one of Britain's most important manufacturers of steam engines show that India was the company's second-largest market after the Netherlands.

Around this time, several steam engines were also built in Calcutta. The skills for the making and maintenance of these machines were abundantly available in and around the city. A historian of Indian steamships notes, "the Ganges valley villages were teeming with men whose skills in their own traditional technology were roughly similar to those needed to keep a steam flotilla in service." This anticipated an important but little-noticed aspect of the age of steam: it was India that provided much of the manpower for the boiler rooms of the world's steam-powered merchant fleets.

Already in the early 1820s, businessmen in India, foreign and local, had become keenly interested in the possibility of a regular steamer service between England and India. Since no coal-fueled vessel had yet made that journey, a conglomerate of wealthy men—a group that included the Nawab of Awadh— announced a prize of ten thousand pounds sterling for the first steamship to complete the voyage in less than seventy days. The challenge was taken up by a group of investors in England and a side-wheeled steamship called *Enterprise* (a name that recurs often among early steamships) was built at Deptford at a cost of forty-three thousand pounds sterling.

The *Enterprise* left Falmouth on August 16, 1825, and arrived in Calcutta on December 7, after a voyage of 114 days. Even though the steamer had not met the specified time limit, the committee decided to award the owners a substantial sum of money in recognition of the historic nature of the journey.

The arrival of the *Enterprise* caused great excitement in Calcutta. In my novel *Flood of Fire*, a character recalls the moment many years later, in Canton:

I well remembered the day, fourteen years ago, when a steamer called *Enterprize* had steamed up to Calcutta . . . this was the first steamer ever to be seen in the Indian Ocean and she had won a prize . . . for her feat. Being young at that time I had expected that *Enterprize* would be a huge, towering vessel: I was astonished to find that she was a small, ungainly-looking craft. But when the *Enterprize* began to move my disappointment had turned to wonder: without a breath of wind stirring, she had gone up and down the Calcutta waterfront, manoeuvring dexterously between throngs of boats and ships.

. . . the arrival of the *Enterprize* had set off a great race amongst the shipowners of Calcutta. Within a few years the New Howrah Dockyards had built the *Forbes*, a teak paddle-wheeler fitted with two sixty-horsepower engines. This had inspired my own father to enter the race: he had invested five thousand rupees in a company launched by the city's most eminent Bengali entrepreneur, Dwarkanath Tagore: it was called the Calcutta Steam Tug Association, and it was soon in possession of two steamers. . . . Steam-tugs are a familiar sight on the Hooghly now; people have grown accustomed to seeing them on the river, churning purposefully through the water and exhaling long trails of smoke, soot and cinders.

Dwarkanath Tagore, whose grandson, the poet Rabindranath Tagore, would win the Nobel Prize for Literature in 1913, is a key figure in the history of India's carbon economy. In the late eighteenth century, at a time when "British investment in Bengal . . . was insignificant," he was one of a number of local businessmen who took the initiative in building a commercial infrastructure. He was also a visionary in regard to the carbon economy. Not only did he set up the Calcutta Steam Tug Association, he was one of the earliest promoters of railways in India. In 1836, Tagore bought the Raniganj coalfields in Bihar, thereby becoming one of the principal suppliers of coal in Bengal. But this venture eventually failed, not because it was inherently flawed, but because it received no support from the East India Company, the ruling power.

On the other side of the subcontinent in Bombay, indigenous merchants were equally enthusiastic about the new technology. In some ways, they were also in a better position to respond to it because Bombay's indigenous shipbuilding industry dated back to the mid-eighteenth century. The Wadia family and their Bombay Dockyard were the industry's leaders, and they were able to compete effectively with the most famous shipyards of Europe and the United States. The Wadias' reputation was such that they obtained many contracts from the Royal Navy (it was their shipyard that built the *Cornwallis*, on which the Treaty of Nanking was signed in 1842).

But it was the very success of Bombay's shipyards that led to their undoing. The English shipping industry complained that "the families of all the shipwrights in England are certain to be reduced to starvation" unless India-built ships were barred from accessing British ports. In 1815, the British Parliament passed a law, the Registry Act, that placed tight restrictions on Indian ships and sailors ("lascars"). It has been said of this law that it was "more devastating to the economy of Indian shipping than all the competitive technological innovations of the last 300 years put together."

It is worth recalling here that shipbuilding was at the leading edge of industrial innovation in the eighteenth and nineteenth centuries; shipbuilders had to respond swiftly to technological advances of all kinds, for both commercial and military reasons. Bombay's shipbuilders were not slow to re-

spond to the challenges of steam technology. In 1830, master builder Naurojee Jamsetjee Wadia launched a steamer, the
Hugh Lindsay, that was fitted with two engines sent out from
England. But he was outdone by his relative, the engineer Ardeseer Cursetjee Wadia, who entered into apprenticeship in
the Bombay Dockyard in 1822 at the age of fourteen and was
eventually elected a Fellow of the Royal Society. Ardeseer wrote
in his memoirs: "My enthusiastic love of science now led me to
construct, unassisted, a small steam engine of about 1 HP. I like-
wise endeavored to explain to my countrymen the nature and
properties of steam; and to effect this point I had constructed
at a great expense in England, a marine steam-engine, which,
being sent out to Bombay, I succeeded with the assistance of
a native blacksmith in fixing in a boat of my own building."

It will be evident from this that Indian entrepreneurs were
quick to grasp the possibilities of British and American steam
technology. There is no reason to suppose that they would not
have been at least as good at imitating it as were their counterparts in, say, Germany or Russia, had the circumstances been
different. It was the very fact that India's ruling power was also
the global pioneer of the carbon economy that ensured that
it could not take hold in India, at that point in time. The appetites of the British economy needed to be fed by large quantities of raw materials, produced by solar-based methods of
agriculture. Had a carbon economy developed synchronously
in India and elsewhere, these materials would have been used
locally instead of being exported.

In other words, the emerging fossil-fuel economies of the
West required that people elsewhere be prevented from developing coal-based energy systems of their own, by compulsion
if necessary. As Timothy Mitchell observes, the coal economy
thus essentially "depended on not being imitated." Imperial
rule assured that it was not.

It was not for any lack of industriousness, then, or ingenuity or entrepreneurial interest, that this avatar of the carbon economy withered in India: the matter might have taken a completely different turn if local industrialists had enjoyed the kind of state patronage that was routinely extended to their competitors elsewhere.

8.

Where it concerns human beings, it is almost always true that the more anxiously we look for purity the more likely we are to come upon admixture and interbreeding. This is no less true, I think, of the genealogy of the carbon economy than it is of the human race: many different lines of descent are commingled in its present form.

The factor that gave the carbon economy its decisive shape was not the provenance of the machines that ushered in the Industrial Revolution: these could have been used and imitated just as easily in other parts of the world as they were in continental Europe. What determined the shape of the global carbon economy was that the major European powers had already established a strong (but by no means hegemonic) military and political presence in much of Asia and Africa at the time when the technology of steam was in its nascency, that is to say, the late eighteenth and early nineteenth centuries. From that point on, carbon-intensive technologies were to have the effect of continually reinforcing Western power with the result that other variants of modernity came to be suppressed, incorporated, and appropriated into what is now a single, dominant model.

The boost that fossil fuels provided to Western power is nowhere more clearly evident than in the First Opium War, where armored steamships, led by the aptly named *Nemesis*, played a decisive role. In other words, carbon emissions were, from very early on, closely co-related to power in all its aspects: this continues to be a major, although unacknowledged, factor in the politics of contemporary global warming.

The Opium War of 1839–42 was the first important conflict to be fought in the name of free trade and unfettered markets; yet, ironically, the most obvious lesson of this period is that capitalist trade and industry cannot thrive without access to military and political power. State interventions have always been critical to its advancement. In Asia, it was military dominance that created the conditions in which Western capital could prevail over indigenous commerce. British imperial officials of that period understood perfectly well the lesson contained in this: it was that the maintenance of military dominance had to be the primary imperative of empire.

In mainland Asia, the crucial linkages between economy, political sovereignty, and military power were not restored till the paired processes of decolonization and the (temporary) retreat of the erstwhile colonial powers were set in motion by the end of the Second World War. It is surely no coincidence that the acceleration of mainland Asian economies followed within a few decades. As Dipesh Chakrabarty points out, the period of the Great Acceleration is precisely "the period of great decolonization in countries that had been dominated by European imperial powers."

Such being the case, another essential question in relation to the chronology of global warming is this: What would have happened if decolonization and the dismantling of empires (including that of Japan) had occurred earlier, say, after the First World War? Would the economies of mainland Asia have accelerated earlier?

If the answer to this were yes, then another, equally important question would arise: Could it be the case that imperialism actually delayed the onset of the climate crisis by retarding the expansion of Asian and African economies? Is it possible that if the major twentieth-century empires had been dismantled earlier, then the landmark figure of 350 parts per million of carbon dioxide in the atmosphere would have been crossed long before it actually was?

It seems to me that the answer is almost certainly yes. This is indeed silently implied in the positions that India, China, and many other nations have taken in global climate negotiations: the argument about fairness in relation to per capita emissions is, in a sense, an argument about lost time.

Here, then, is the paradoxical possibility that is implied by these positions: the fact that some of the key technologies of the carbon economy were first adopted in England, the world's leading colonial power, *may actually have retarded the onset of the climate crisis.*

To acknowledge the complexity of the history of the carbon economy is not in any way to diminish the force of the argument for global justice regarding greenhouse gas emissions. To the contrary, it places that argument within the same contexts as debates about inequality, poverty, and social justice within countries like Britain and the United States: it is to assert that the poor nations of the world are not poor because they were indolent or unwilling; their poverty is itself an effect of the inequities created by the carbon economy; it is the result of systems that were set up by brute force to ensure that poor nations remained always at a disadvantage in terms of both wealth and power.

Inasmuch as the fruits of the carbon economy constitute wealth, and inasmuch as the poor of the global south have historically been deprived of this wealth, it is certainly true, by every available canon of distributive justice, that they are entitled to a greater share of the rewards of that economy. But even to enter into that argument is to recognize how deeply we are mired in the Great Derangement: our lives and our choices are enframed in a pattern of history that seems to leave us nowhere to turn but toward our self-annihilation.

"Money flows toward short term gain," writes the geologist David Archer, "and toward the over-exploitation of unregulated common resources. These tendencies are like the invisible hand of fate, guiding the hero in a Greek tragedy toward his inevitable doom."

This is indeed the essence of humanity's present derangement.

9.

Imperialism was not, however, the only obstacle in Asia's path to industrialization: this model of economy also met with powerful indigenous resistances of many different kinds. While it is true that industrial capitalism met with resistance on every continent, not least Europe, what is distinctive in the case of Asia is that the resistance was often articulated and championed by figures of extraordinary moral and political authority, such as Mahatma Gandhi. Among Gandhi's best-known pronouncements on industrial capitalism are these famous lines written in 1928: "God forbid that India should ever take to industrialism after the manner of the West. If an entire nation of 300 millions [sic] took to similar economic exploitation, it would strip the world bare like locusts."

This quote is striking because of the directness with which it goes to the heart of the matter: numbers. It is proof that Gandhi, like many others, understood intuitively what Asia's history would eventually demonstrate: that the universalist premise of industrial civilization was a hoax; that a consumerist mode of existence, if adopted by a sufficient number of people, would quickly become unsustainable and would lead, literally, to the devouring of the planet.

Of course, Gandhi was not alone in being granted this insight; many others around the world were to arrive at the same conclusion, often by completely different routes. But Gandhi occupied a position of unique social and cultural importance, and, what was more, he was willing to carry his vision to its logical conclusion by voluntarily renouncing, on behalf of his nation, the kind of power and affluence that is conferred by industrial civilization.

This was perfectly well understood by Gandhi's political enemies on the Hindu right, who insistently characterized him as a man who wanted to weaken India. And indeed it was for this very reason that Gandhi was assassinated by a former member of an organization that would later become the nucleus of the political formation that now rules India. This coalition came to power by promising exactly what Gandhi had renounced: endless industrial growth.

In China, similarly, as Prasenjit Duara has shown, industrialism and consumerism faced powerful resistances from within the Taoist, Confucian, and Buddhist traditions. There too many influential thinkers understood the implications of large-scale modernization. One such was Zhang Shizhao (1881–1973) who was minister of education in Duan Qirui's government: "While finitude characterizes all things under heaven," he wrote, "appetites alone know no bounds. When the amount of what is of finite supply is gauged on the basis of boundless appetites, the exhaustion of the former can be expected within a matter of days. Conversely, the depletion of finite things would soon come when used to satisfy insatiable desires."

Duara has shown in rich detail how the resistance to capitalist modernity was overcome very slowly in both of Asia's most populous countries, through a range of political and cultural

movements that would lead, over time, to "the Protestantiza-
tion of religions, secularization . . . and nation-building."

But the Asian countries that industrialized first did not, in
fact, follow the Western model: as Sugihara and others have
shown, the path that Japan and Korea took was, of necessity,
much less wasteful of resources. Japan diverged from the West
in another way as well: an awareness of natural constraints became a part of its official ideology, which insisted that "nature
is consciousness for the Japanese people."

It is a striking fact also that many leading figures from Asia
voiced concerns even at a time when environmentalism was
largely a countercultural issue in the West. One of them was
the Burmese statesman U Thant, who served as the secretary-
general of the United Nations from 1962 to 1971 and was in-
strumental in establishing the United Nations Environment
Programme. In 1971, he issued a warning that seems strangely
prescient today: "As we watch the sun go down, evening after
evening, through the smog across the poisoned waters of our
native earth, we must ask ourselves seriously whether we really wish some future universal historian on another planet
to say about us: 'With all their genius and with all their skill,
they ran out of foresight and air and food and water and ideas,'
or, 'They went on playing politics until their world collapsed
around them.'"

In China, an awareness of the importance of numbers would
lead eventually to the recently ended One-Child Policy, a measure that, at the cost of inflicting great suffering, has had the
effect of stabilizing the country's population at a level far below
what it might otherwise have been. Draconian and repressive
as this policy undoubtedly was, from the reversed perspective
of the Anthropocene it may one day be claimed as a mitigatory measure of great significance. For if it is indeed the case
that the onset of the climate crisis has been accelerated by

the industrialization of mainland Asia, then we may be sure that with several hundred million more consumers included in the equation the landmark figure of 350 parts per million of carbon dioxide in the atmosphere would have been passed very much earlier.

In any reckoning of climate justice, this history too needs to be taken into account: that in both India and China, the two nations that are now often blamed for precipitating the climate crisis, there were significant numbers of people who understood, long before climate scientists brought in the data, that industrial civilization was subject to limitations of scale and would collapse if adopted by the majority of the earth's people. Although they may finally have failed to lead their compatriots in a different direction, they did succeed in retarding the wholesale adoption of a consumerist, industrial model of economy in their countries. In a world where the rewards of a carbon-intensive economy are regarded as wealth, this must be reckoned as a very significant material sacrifice, for which they can, quite legitimately, demand recognition.

The demand for "climate reparations" is therefore founded on unshakeable grounds, historically and ethically. Yet the complexity of the carbon economy's genealogy holds a lesson also for those in the global south who would draw a wide and clear line between "us" and "them" in relation to global warming. While there can be no doubt that the climate crisis was brought on by the way in which the carbon economy evolved in the West, it is also true that the matter might have taken many different turns. The climate crisis cannot therefore be thought of as a problem created by an utterly distant "Other."

The phrase "common but differentiated responsibilities," frequently heard during the Paris climate change negotiations of 2015, is thus a rare example of bureaucratese that is both apt and accurate. Anthropogenic climate change, as Chakrabarty and others have pointed out, is the unintended consequence of the very existence of human beings as a species. Although different groups of people have contributed to it in vastly different measure, global warming is ultimately the product of the totality of human actions over time. Every human being who has ever lived has played a part in making us the dominant species on this planet, and in this sense every human being, past and present, has contributed to the present cycle of climate change.

The events of today's changing climate, in that they represent the totality of human actions over time, represent also the terminus of history. For if the entirety of our past is contained within the present, then temporality itself is drained of significance. Or, in the words of the Japanese philosopher Watsuji Tetsuro: "Rather than trace historical development ... all one need do is to distinguish the various formal transformations of the present."

The climate events of this era, then, are distillations of all of human history: they express the entirety of our being over time.

Politics

1.

Climate change poses a powerful challenge to what is perhaps the single most important political conception of the modern era: the idea of freedom, which is central not only to contemporary politics but also to the humanities, the arts, and literature.

Since the Enlightenment, as Dipesh Chakrabarty has pointed out, philosophers of freedom were "mainly, and understandably, concerned with how humans would escape the injustice, oppression, inequality, or even uniformity foisted on them by other humans or human-made systems." Nonhuman forces and systems had no place in this calculus of liberty: indeed being independent of Nature was considered one of the defining characteristics of freedom itself. Only those peoples who had thrown off the shackles of their environment were thought to be endowed with historical agency; they alone were believed to merit the attention of historians—other peoples might have had a past but they were thought to lack history, which realizes itself through human agency.

Now that the stirrings of the earth have forced us to recognize that we have never been free of nonhuman constraints how are we to rethink those conceptions of history and agency? The same question could be posed with equal force in relation to art and literature, particularly in regard to the twentieth century, when there was a radical turn away from the nonhuman to the human, from the figurative toward the abstract.

These developments were not, of course, generated by purely aesthetic considerations. They were influenced also by politics, especially the politics of the Cold War—as, for example, when American intelligence agencies intervened to promote abstract expressionism against the social realism favored by the USSR.

But the trajectory of the arts had been determined long before the Cold War: through the twentieth century they followed a course that led them to become increasingly self-reflexive. "Twentieth-century art," wrote Roger Shattuck in 1968, "has tended to *search itself* rather than exterior reality for beauty of meaning or truth, a condition that entails a new relationship between the work of art, the world, the spectator, and the artist." It was thus that human consciousness, agency, and identity came to be placed at the center of every kind of aesthetic enterprise.

In this realm, too, Asia has played a special role: the questions that animated, obsessed, and haunted the thinkers and writers of twentieth-century Asia were precisely those that related to the "modern." Jawaharlal Nehru's passion for dams and factories and Mao Zedong's "War on Nature" had their counterparts also in literature and the arts.

In their embrace of modernity, Asian writers and artists created ruptures that radically reconfigured the region's literature, art, architecture, and so on. In Asia as elsewhere, this meant that the abstract and the formal gained ascendancy over the figurative and the iconographic; it meant also that many traditions, including those that accorded the nonhuman a special salience, were jettisoned. Here, as elsewhere, freedom came to be seen as a way of "transcending" the constraints of material life—of exploring new regions of the human mind, spirit, emotion, consciousness, interiority: freedom became a quantity that resided entirely in the minds, bodies, and desires of human beings. There is, of course, as Moretti notes, a sort of "ascetic heroism" in such a vision, but it is also clear now that the more "radical and clear-sighted the aesthetic achievements of that time, the more unliveable the world [they] depict."

And now, when we look back upon that time, with our gaze reversed, having woken against our will to the knowledge that we have always been watched and judged by other eyes, what stands out? Is it possible that the arts and literature of this time will one day be remembered not for their daring, nor for their championing of freedom, but rather because of their complicity in the Great Derangement? Could it be said that the "stance of unyielding rage against the official order" that the artists and writers of this period adopted was actually, from the perspective of the Anthropocene, a form of collusion? Recent years have certainly demonstrated the truth of an observation that Guy Debord made long ago: that spectacular forms of rebelliousness are not, by any means, incompatible with a "smug acceptance of what exists . . . for the simple reason that dissatisfaction itself becomes a commodity."

If such a judgement—or even the possibility of it—seems shocking, it is because we have come to accept that the front ranks of the arts are in some way in advance of mainstream culture; that artists and writers are able to look ahead, not just in aesthetic matters, but also in regard to public affairs. Writers and artists have themselves embraced this role with increasing fervor through the twentieth century, and never more so than in the period in which carbon emissions were accelerating.

As proof of this, let us imagine for a moment, just as a thought experiment, that a graph could be drawn of the political engagements of writers and artists through the twentieth century and into the twenty-first. It is quite likely, I suspect, that such a graph would closely resemble a chart of greenhouse gas emissions over the same period: that is to say, the line would indicate a steep and steady rise over the decades, with a few sudden and dramatic upsurges. The First World War would represent one such escalation, the rise in industrial and military activity being mirrored by an enormous outpouring of literature, much of it explicitly political.

During the interwar years, too, the graphs would remain on roughly parallel tracks, a rise in worldwide industrial activity being matched by the increasingly visible involvement of writers with political movements, such as socialism, communism, antifascism, nationalism, and anti-imperialism: Lorca, Brecht, Orwell, Lu Xun, and Tagore being cases in point.

Only in the early post–Second World War decades would there be a marked divergence in the two graphs, with the political engagements of writers outpacing the rise in the rate of emissions. The large-scale industrialization of Asia had yet to begin, after all, while writers around the world were broadening their political engagements on every front. We need think only of the Progressive Writers Movement in India and Pakistan; of decolonization and Sartre; James Baldwin and the civil rights movement; the Beats and the student uprisings of the 1960s; the persecution of Pramoedya Ananta Toer in Indonesia and of Solzhenitsyn in the Soviet Union. This was a time when writers were in the forefront of every political movement around the world.

Not till the 1980s would the graphs again converge, and then, too, not because of any diminution in the political energies of writers and artists but only because the rate of emissions from Asia had begun its steady upward climb. But in this period too, writers were in the vanguard of many movements, feminism and gay rights being but two of them. This was also a time in which the paradoxical coupling of the processes of decolonization, on the one hand, and the increasing hegemony of the English language, on the other, made it possible for writers like myself to enter the global literary mainstream in a way that had not been possible in the preceding two centuries. At the same time, changes in technologies of communication, and a rapid growth in networks of translation, served to internationalize both politics and literature to a point where it could be said that Goethe's vision of a "world literature" (*Weltliteratur*) had come close to being realized.

I can attest from my own experience that this period— when an exploding rate of carbon emissions was rewriting the planet's destiny—was a breathtakingly exciting time in which to launch upon a career as a writer. As I've noted before, not the least aspect of this was the promise of "being ahead" (*en avant*, of being a part, in effect, of an avant-garde), and this conception has been one of the animating forces of the literary and artistic imagination since the start of the twentieth century. "Modernism wrote into its scripture a major text," goes Roger Shattuck's wry observation, "the avant-garde we have with us always."

To want to be ahead, and to celebrate and mythify this endeavor, is indeed one of the most powerful impulses of modernity itself. If Bruno Latour is right, then to be modern is to envision time as irreversible, to think of it as a progression that is forever propelled forward by revolutionary ruptures: these in turn are conceived of on the analogy of scientific innovations, each of which is thought to render its predecessor obsolete.

And obsolescence is indeed modernity's equivalent of perdition and hellfire. That is why this era's most potent words of damnation, passed down in an unbroken relay from Hegel and Marx to President Obama, is the malediction of being "on the wrong side of history."

That the world's most powerful leader should hurl these words at his enemies, in much the same way that curses and imprecations were once used by kings, priests, and shamans, is of course a disavowal of the very irreversibility of time that the mantra invokes: for is it not also an acknowledgment of the power that words have possessed through the ages, of striking fear into the hearts of foes, of conjuring up visions of terror with curses and maledictions? And for modern man, terror is exactly what is evoked by the fear of being left behind, of being "backward."

There is perhaps no better means of tracking the diffusion of modernity across the globe than by charting the widening grip of this fear, which was nowhere more powerfully felt than in the places that were most visibly marked by the stigmata of "backwardness." It was what drove artists and writers in Asia, Africa, and the Arab world to go to extraordinary lengths to "keep up" with each iteration of modernity in the arts: surrealism, existentialism, and so on. And far from diminishing over time, the impulse gathered strength through the twentieth century, so that writers of my generation were, if anything, even less resistant to its power than were our predecessors: we could not but be aware of the many "isms"—structuralism, postmodernism, postcolonialism—that flashed past our eyes with ever-increasing speed.

This is why it comes as a surprise—a shock, really—to look back upon that period of surging carbon emissions and recognize that very few (and I do not exempt myself from this) of the literary minds of that intensely *engagé* period were alive to the archaic voice whose rumblings, once familiar, had now become inaudible to humanity: that of the earth and its atmosphere.

I do not mean to imply that there were no manifestations of a general sense of anxiety and foreboding in the literature of that time; nor do I mean to suggest that mankind had ceased to be haunted by intuitions of apocalypse. These were certainly no less abundant in the last few decades than they have been since stories were first told. It is when I try to think of writers whose imaginative work communicated a more specific sense of the accelerating changes in our environment that I find myself at a loss; of literary novelists writing in English only a handful of names come to mind: J. G. Ballard, Margaret Atwood, Kurt

Vonnegut Jr., Barbara Kingsolver, Doris Lessing, Cormac Mc-
Carthy, Ian McEwan, and T. Coraghessan Boyle. No doubt many
other names could be added to this list, but even if it were to
be expanded a hundredfold or more, it would remain true, I
think, that the literary mainstream, even as it was becoming
more *engagé* on many fronts, remained just as unaware of the
crisis on our doorstep as the population at large.

In this regard, the avant-garde, far from being "ahead," was
clearly a laggard. Could it be, then, that the same process that
inaugurated the rising death spiral of carbon emissions also
ensured, in an uncannily clever gesture of self-protection, that
the artists, writers, and poets of that era would go racing off
in directions that actually blinded them to exactly what they
thought they were seeing: that is to say, what lay *en avant*, what
was to come? And if this were so, would it not be a damning
indictment of a vision in which the arts are seen to be moving
forever forward, in a dimension of irreversible time, by means
of innovation and the free pursuit of imagination?

2.

Writers are not alone, of course, in having broadened and intensified their political and social engagements over the last
couple of decades: this has happened to the entirety of what
used to be called "the intelligentsia." In no small part has this
been brought about by changes in the technology of communication: the Internet and the digital media have made the
sphere of the political broader and more intrusive than ever
before. Today everybody with a computer and a web connection is an activist. Yet what I said earlier about literary circles is
true also of the intelligentsia, and indeed of circles far beyond:
generally speaking, politicization has not translated into a
wider engagement with the crisis of climate change.

The lack of a transitive connection between political mo-
bilization, on the one hand, and global warming, on the other,
is nowhere more evident than in the countries of South Asia,
all of which are extraordinarily vulnerable to climate change.
In the last few decades, India has become very highly politi-
cized; great numbers take to the streets to express indignation
and outrage over a wide range of issues; on television channels and social media, people speak their minds ever more
stridently. Yet climate change has not resulted in an outpouring of passion in the country. This despite the fact that India
has innumerable environmental organizations and grassroots
movements. The voices of the country's many eminent climate
scientists, environmental activists, and reporters do not appear
to have made much of a mark either.

What is true of India is true also of Pakistan, Bangladesh,
Sri Lanka, and Nepal: climate change has not been a significant political issue in any of those countries, even though the
impacts are already being felt across the Indian subcontinent,
not only in an increasing number of large-scale disasters but
also in the form of a slow calamity that is quietly but inexo-
rably destroying livelihoods and stoking social and political
conflicts. Instead, political energy has increasingly come to be
focused on issues that relate, in one way or another, to questions of identity: religion, caste, ethnicity, language, gender
rights, and so on.

The divergence between the common interest and the pre-
occupations of the public sphere points to a change in the
nature of politics itself. The political is no longer about the
commonweal or the "body politic" and the making of collective decisions. It is about something else.

What, then, is that "something"?

A similar question could be posed in relation to the literary imaginary: Why is it increasingly open to certain concep-

tions of the political while remaining closed to an issue that concerns our collective survival?

Here again the trajectory of the modern novel represents, I think, a special case of a broader cultural phenomenon. The essence of this phenomenon is again captured by the words that John Updike used to characterize the modern novel: "individual moral adventure." I have already addressed one of the implications of this conception of the novel: the manner in which it banishes the collective from the territory of the fictional imagination. I want to attend now to another aspect of it: the implications of the word *moral*.

We encounter this word very frequently today in relation to fiction as well as politics. In my view, the notion of "the moral" is the hinge that has made possible the joining of the political and the literary imaginary.

The word *moral* derives from a Latin root signifying "custom" or "mores"; connotations of aristocratic usages may well, as Nietzsche famously argued, have been implicit in it. The word has had a long career in English: having once resided within the Church—especially the churches of Protestantism—it has now come to draw its force primarily from the domain of the political. But this is not a politics that is principally concerned with the ordering of public affairs. It is rather a politics that is also increasingly conceived of as an "individual moral adventure" in the sense of being an interior journey guided by the conscience. Just as novels have come to be seen as narratives of identity, so too has politics become, for many, a search for personal authenticity, a journey of self-discovery.

Although the evolution of the term moral has brought it squarely into the secular domain, the term continues to be powerfully marked by its origins, which clearly lie within Christianity and particularly Protestantism. The moral-political, as thus conceived, is essentially Protestantism without a God: it commits its votaries to believing in perfectibility, individual redemption, and a never-ending journey to a shining city on a hill—constructed, in this instance, not by a deity, but by democracy. This is a vision of the world as a secular church, where all the congregants offer testimony about their journeys of self-discovery.

This imagining of the world has profound consequences for fiction as well as the body politic. Fiction, for one, comes to be reimagined in such a way that it becomes a form of bearing witness, of testifying, and of charting the career of the conscience. Thus do sincerity and authenticity become, in politics as in literature, the greatest of virtues. No wonder, then, that one of the literary icons of our age, the novelist Karl Ove Knausgaard, has publicly admitted to "being sick of fiction." As opposed to the "falsity" of fiction, Knausgaard has "set out to write exclusively from his own life." This is not, however, a new project: it belongs squarely within the tradition of "diary keeping and spiritual soul-searching [that] . . . was a central aspect of Puritan religiosity." This secular baring-of-the-soul is exactly what is demanded by the world-as-church.

If literature is conceived of as the expression of authentic experience, then fiction will inevitably come to be seen as "false." But to reproduce the world as it exists need not be the project of fiction; what fiction—and by this I mean not only the novel but also epic and myth—makes possible is to approach the world in a subjunctive mode, to conceive of it *as if* it were other than it is: in short, the great, irreplaceable potentiality of fiction is that it makes possible the imagining of possibilities. And to imagine other forms of human existence is exactly the challenge that is posed by the climate crisis: for if there is any one thing that global warming has made perfectly clear it is that to think about the world only as it is amounts to a formula for collective suicide. We need, rather, to envision what it might be. But as with much else that is uncanny about the Anthropocene, this challenge has appeared before us at the very moment when the form of imagining that is best suited to answering it—fiction—has turned in a radically different direction.

This then is the paradox and the price of conceiving of fiction and politics in terms of individual moral adventures: it negates possibility itself. As for the nonhuman, it is almost by definition excluded from a politics that sanctifies subjectivity and in which political claims are made in the first person. Consider, for example, the stories that congeal around questions like, "Where were you when the Berlin Wall fell?" or "Where were you on 9/11?" Will it ever be possible to ask, in the same vein, "Where were you at 400 ppm [parts per million]?" or "Where were you when the Larsen B ice shelf broke up?"

For the body politic, this vision of politics as moral journey has also had the consequence of creating an ever-growing divergence between a public sphere of political performance and the realm of actual governance: the latter is now controlled by largely invisible establishments that are guided by imperatives of their own. And as the public sphere grows ever more performative, at every level from presidential campaigns to online petitions, its ability to influence the actual exercise of power becomes increasingly attenuated.

This was starkly evident in the buildup to the Iraq War in 2003: I was in New York on February 15 that year, and I joined the massive antiwar demonstration that wound through the avenues of mid-Manhattan. Similar demonstrations were staged in six hundred other cities, in sixty countries around the world; tens of millions of people took part in them, making them possibly the largest single manifestation of public dissent in history. Yet even at that time there was a feeling of hopelessness; relatively few, I suspect, believed that the marches would effect a change in policy—and indeed they did not. Then, as never before, it became clear that the public sphere's ability to influence the security and policy establishment had eroded drastically.

Since then the process has only accelerated: in many other matters, like austerity, surveillance, drone warfare, and so on, it is now perfectly clear that in the West political processes exert very limited influence over the domain of statecraft— so much so that it has even been suggested that "citizens no longer *seriously expect* . . . that politicians will really represent their interests and implement their demands."

This altered political reality may in part be an effect of the dominance of petroleum in the world economy. As Timothy Mitchell has shown, the flow of oil is radically unlike the movement of coal. The nature of coal, as a material, is such that its transportation creates multiple choke points where organized labor can exert pressure on corporations and the state. This is not case with oil, which flows through pipelines that can bypass concentrations of labor. This was exactly why British and American political elites began to encourage the use of oil over coal after the First World War.

These efforts succeeded perhaps beyond their own wildest dreams. As an instrument of disempowerment oil has been spectacularly effective in removing the levers of power from the reach of the populace. "No matter how many people take to the streets in massive marches," writes Roy Scranton, "they cannot put their hands on the real flows of power because they do not help to produce it. They only consume."

Under these circumstances, a march or a demonstration of popular feeling amounts to "little more than an orgy of democratic emotion, an activist-themed street fair, a real-world analogue to Twitter hashtag campaigns: something that gives you a nice feeling, says you belong in a certain group, and is completely divorced from actual legislation and governance."

In other words, the public sphere, where politics is performed, has been largely emptied of content in terms of the exercise of power: as with fiction, it has become a forum for secular testimony, a baring-of-the-soul in the world-as-church. Politics as thus practiced is primarily an exercise in personal expressiveness. Contemporary culture in all its aspects (including religious fundamentalisms of almost every variety) is pervaded by this expressivism, which is itself "to a significant degree a result of the strong role of Protestant Christianity in the making of the modern world." There could be no better vehicle for this expressivism than the Internet, which makes various means of self-expression instantly available through social media. And as tweets and posts and clips circle the globe, they generate their mirror images of counterexpression in a dynamic that quickly becomes a double helix of negation.

As far back as the 1960s Guy Debord argued in his seminal book *The Society of the Spectacle*: "The whole life of those societies in which modern conditions of production prevail presents itself as an immense accumulation of spectacles. All that was once directly lived has become mere representation." The ways in which political engagements unfold over social media confirm this thesis, propounded long before the Internet became so large a part of our lives: "The spectacle is by definition immune from human activity, inaccessible to any projected review or correction. *It is the opposite of dialogue.* Wherever representation takes on an independent existence, the spectacle reestablishes its rule."

The net result is a deadlocked public sphere, with the actual exercise of power being relegated to the interlocking complex of corporations and institutions of governance that has come to be known as the "deep state." From the point of view of corporations and other establishment entities, a deadlocked public is, of course, the best possible outcome, which, no doubt, is why they frequently strive to produce it: the funding of climate change "denial" in the United States and elsewhere, by corporations like Exxon—which have long known about the consequences of carbon emissions—is a perfect example of this.

In effect, the countries of the West are now in many senses "post-political spaces" that are managed by apparatuses of various kinds. For many, this creates a haunting sense of loss that manifests itself in an ever-more-desperate yearning to recoup a genuinely participatory politics. This is in no small part the driving force behind such disparate figures as Jeremy Corbyn and Bernie Sanders, on the one hand, and Donald Trump, on the other. But the collapse of political alternatives, the accompanying disempowerment, and the ever-growing intrusion of the market have also produced responses of another kind—nihilistic forms of extremism that employ methods of spectacular violence. This too has taken on a life of its own.

3.

The public politics of climate change is itself an illustration of the ways in which the moral-political can produce paralysis.

Of late, many activists and concerned people have begun to frame climate change as a "moral issue." This has become almost a plea of last resort, appeals of many other kinds having failed to produce concerted action on climate change. So, in an ironic twist, the individual conscience is now increasingly seen as the battleground of choice for a conflict that is self-evidently a problem of the global commons, requiring collective action: it is as if every other resource of democratic governance had been exhausted leaving only this residue—the moral.

This framing of the issue certainly has one great virtue, in that it breaks decisively with the economistic, cost-benefit language that the international climate change bureaucracy has imposed on it. But at the same time, this approach also invokes a "politics of sincerity" that may ultimately work to the advantage of those on the opposite side. For if the crisis of climate change is to be principally seen in terms of the questions it poses to the individual conscience, then sincerity and consistency will inevitably become the touchstones by which political positions will be judged. This in turn will enable "deniers" to accuse activists of personal hypocrisy by pointing to their individual lifestyle choices. When framed in this way, authenticity and sacrifice become central to the issue, which then comes to rest on matters like the number of lightbulbs in Al Gore's home and the forms of transport that demonstrators use to get to a march.

I saw a particularly telling example of this in a TV interview with a prominent activist after the New York climate change march of September 2014. The interviewer's posture was like that of a priest interrogating a wayward parishioner; her questions were along the lines of "What have *you* given up for climate change? What are *your* sacrifices?"

The activist in question was quickly reduced to indignant incoherence. So paralyzing is the effect of the fusion of the political and the moral that he could not bring himself to state the obvious: that the scale of climate change is such that individual choices will make little difference unless certain collective decisions are taken and acted upon. Sincerity has nothing to do with rationing water during a drought, as in today's California: this is not a measure that can be left to the individual conscience. To think in those terms is to accept neo-liberal premises.

Second, yardsticks of morality are not the same everywhere. In many parts of the world, and especially in English-speaking countries, canons of judgment on many issues still rest on that distinctive fusion of economic, religious, and philosophical conceptions that was brought about by the Scottish Enlightenment. The central tenet of this set of ideas, as John Maynard Keynes once put it, is that "by the working of natural laws individuals pursuing their own interests with enlightenment, in condition of freedom, always tend to promote the general interest at the same time!"

The "everyday political philosophy of the nineteenth century" (as Keynes described it) remains an immensely powerful force in the United States and elsewhere: for those on the right of the political spectrum, this set of ideas retains something of its millenarian character with individualism, free trade, and God constituting parts of a whole. But by no means is it only the religiously minded whose ideas are shaped by this philosophy: it is worth noting that the dominant secular paradigms of ethics in the United States—for example, as in John Rawls's theory of justice—are also founded upon assumptions about individual rationality that are borrowed from neoclassical economics.

It is instructive in this regard to look at an area of the humanities that has been unusually quick to respond to climate change: the subdiscipline of philosophy represented by climate ethicists. The dominant approach in this discipline is again posited on rational actors, freely pursuing their own interests. A philosopher of this tradition, in responding to the argument that the moral imperative of climate change comes from the need to save the millions of lives in Asia, Africa, and elsewhere, might well quote David Hume: "'Tis not contrary to reason to prefer the destruction of the whole world to the scratching of my finger." Climate activists' appeals to morality will not necessarily find much support here.

Last, we already know, from the example of Mahatma Gandhi, that the industrial, carbon-intensive economy cannot be fought by a politics of sincerity. Gandhi invested himself, body and soul, in the effort to prevent India from adopting the Western, industrial model of economy. Drawing on many different traditions, he articulated and embodied a powerful vision of renunciatory politics; no reporter would have had the gall to ask him what he had sacrificed; his entire political career was based upon the idea of sacrifice. Gandhi was the very exemplar of a politics of moral sincerity.

Yet, while Gandhi may have succeeded in dislodging the British from India, he failed in this other endeavor, that of steering India along a different economic path. He was able, at best, to slightly delay a headlong rush toward an all-devouring, carbon-intensive economy. There is little reason to believe that a politics of this kind will succeed in relation to global warming today.

Climate change is often described as a "wicked problem." One of its wickedest aspects is that it may require us to abandon some of our most treasured ideas about political virtue: for example, "be the change you want to see." What we need instead is to find a way out of the individualizing imaginary in which we are trapped.

When future generations look back upon the Great Derangement they will certainly blame the leaders and politicians of this time for their failure to address the climate crisis. But they may well hold artists and writers to be equally culpable—for the imagining of possibilities is not, after all, the job of politicians and bureaucrats.

4.

One of the most important factors in the global politics of climate change is the role the Anglosphere plays in today's world. This is true for many reasons, not the least of which is that the

Anglosphere is no longer a notional entity: it has been given formal expression in the Five Eyes alliance that now binds the intelligence and surveillance structures of the United States, Great Britain, Australia, Canada, and New Zealand. The UKUSA Security Agreement that formalized the arrangement implicitly acknowledges that this alliance undergirds the world's current security architecture.

The fact that laissez-faire ideas are still dominant within the Anglosphere is therefore itself central to the climate crisis. In that global warming poses a powerful challenge to the idea that the free pursuit of individual interests always leads to the general good, it also challenges a set of beliefs that underlies a deeply rooted cultural identity, one that has enjoyed unparalleled success over the last two centuries. Much of the resistance to climate science comes exactly from this, which is probably why the rates of climate change denial tend to be unusually high throughout the Anglosphere.

Yet it is also true that the Anglosphere, the United States in particular, has produced the overwhelming bulk of climate science, as well as some of the earliest warnings of global warming. Moreover, many, if not most, of those who have taken the lead on the issue politically, whether it be as thinkers, theorists, or activists, are from these five countries, which together possess some of the most vigorous environmental movements in the world. Bill McKibben's 350.org is but one example of a group that has spearheaded a global movement.

The tension between these two polarities—widespread denialism, on the one hand, and vigorous activist movements, on the other—now defines the public politics of climate change throughout the Anglosphere, but particularly in the United States. And since identity and performativity are now central to public discourse, climate change too has become enmeshed with the politics of self-definition. When American and Australian politicians speak of climate change negotiations as posing a threat to "our way of life," they are following the same script that led Ronald Reagan to speak of the reduction of the use of oil as an assault on what it means to be American.

The enmeshment of global warming with issues of an entirely different order has given a distinctive turn to the politics of climate change in the Anglosphere. Instead of being seen as a phenomenon that requires a practical response, as it largely is in Holland and Denmark, or as an existential danger, as it is in the Maldives and Bangladesh, it has become one of many issues that are clustered along a fault line of extreme political polarization. Those on the rightward side of this line view climate science through a conspiratorial lens, linking it with socialism, communism, and so on. (As Naomi Oreskes and Erik Conway have noted, some of the most influential scientific denialists may have been motivated by the ideology of the Cold War.) These associations have, in turn, generated an extraordinary degree of rancor toward some climate scientists, some of whom, like Michael E. Mann, have had to face all manner of threats, harassment, and intimidation. It is a tribute to their courage that they have persevered with their work despite these attacks.

The opposition to climate science is not, however, a self-subsisting phenomenon. As Oreskes, Conway, and others have shown, it is enabled, encouraged, and funded by certain corporations and energy billionaires. These vested interests have supported organizations that systematically spread misinformation and create confusion within the electorate. The situation is further compounded by the mass media, which have generally underplayed climate change and have sometimes even distorted the findings of climate scientists. This bias owes much, no doubt, to the fact that large sections of the media are now controlled by climate skeptics like Rupert Murdoch, and by corporations that have vested interests in the carbon economy. The net result, in any case, is that the denial and disputing of scientific findings has become a major factor in the climate politics of the Anglosphere.

Yet I think it would be a mistake to assume that denialism within the Anglosphere is only a function of money and manipulation. There is an excess to denialist attitudes that suggests that the climate crisis threatens to unravel something deeper, without which large numbers of people would be at a loss to find meaning in their history and indeed their existence in the world.

In other words, the climate crisis has given the lie to Max Weber's contention that modernity brings about the disenchantment of the world. Bruno Latour has long argued that this disenchantment never happened and this is now plain for all to see. The "everyday political philosophy of the nineteenth century" is, as Keynes understood very well, an enchantment just as powerful as any dithyrambic mythology. And it is perhaps even harder to disavow because it comes disguised as a truthful description of the world; as fact, not fantasy. This perhaps is why, despite every effort to disseminate accurate information about climate science, the public domain of the Anglosphere remains deeply divided on the issue of climate change.

But strangely, the picture takes on a completely different appearance when we look to other domains of the American body politic, for example, the security establishment. There is no sign there of either denial or confusion: to the contrary, the Pentagon devotes more resources to the study of climate change than any other branch of the U.S. government. The writer and climate activist George Marshall notes, "the most rational and considered response to the uncertainties of climate change can be found among military strategists. . . . As General Chuck Wald, former deputy commander of U.S. European Command puts it: 'There's a problem there and the military is going to be a part of the solution.'" Other top-ranking officers have been equally blunt. In 2013, when Admiral Samuel J. Locklear III (then head of the U.S. Pacific Command) was asked about the "biggest long-term security threat to the United States in the Pacific Region," he pointed immediately to climate change, identifying it as the factor that was most likely to "cripple the security environment."

Indeed, the U.S. military establishment's focus on global warming is such that Col. Lawrence Wilkerson, former chief of staff to Secretary of State Colin Powell, once summed it up with these words: "The only department in . . . Washington that is clearly and completely seized with the idea that climate change is real is the Dept. of Defense."

The seriousness of this commitment is evident in the fact that the U.S. military—which is also the single largest user of fossil fuels in the country—has launched several hundred renewable energy initiatives and is investing heavily in biofuels, microgrids, electric vehicles, and so on. Between 2006 and 2009, its investments in this sector rose by 200 percent, to over a billion dollars, and is expected to go up to $10 billion by 2030. All of this has been done in such a way as to bypass the contentious debates of the public sphere.

Indeed, it would seem that the American military has in some instances appropriated the language and even the tactics of climate change activism. "Not only has the grand narrative of climate change been co-opted, warped and re-routed by the proponents of *green security*," write Sanjay Chaturvedi and Timothy Doyle, "the very forms of new social movement resistance have been copied and reworked to suit these most recent geopolitical moments. In these multi-layered, multidirectional spaces, neo-liberal economics and neo-securities are one."

Similarly, U.S. intelligence agencies, and personnel associated with them, have produced some of the earliest and most detailed studies of the security implications of climate change. In 2013, James Clapper, the highest-ranking intelligence official in the United States, testified to the Senate that "extreme weather events (floods, droughts, heat waves) will increasingly disrupt food and energy markets, exacerbating state weakness, forcing human migrations, and triggering riots, civil disobedience, and vandalism."

In addition, American intelligence services have already made the surveillance of environmentalists and climate activists a top priority. This has been greatly facilitated, on the one hand, by the widening powers granted to security agencies in the "permanent state of emergency" of the post-9/11 era, and, on the other, by the increasing privatization of intelligence gathering in recent years. The latter development has led to the emergence of a "gray intelligence" industry through a "blurring of public and private spying," and this in turn has made it possible for corporations as well government agencies to infiltrate and spy on environmental groups of many different kinds.

In short, in the United States climate activists are now among the prime targets of a rapidly growing surveillance-industrial complex. This would hardly be the case if the vast American intelligence establishment were in denial about the reality of climate change.

The British military posture is similar; this is how a report by an Australian military think tank sums it up: "From mainstreaming climate change into national planning to appointing senior military authorities to lead on climate change within the defence force, the UK and US governments have directed their militaries to rapidly prepare for climate change and its impacts." The Australian defense establishment is also working hard to coordinate its climate security strategy with the United States and United Kingdom: this posture has been maintained even at times when the stance of the country's political leadership was denialist.

5.

Clearly, despite the deep public divisions in the Anglosphere, there is no denial or division about global warming within the military and intelligence establishments of these countries: to the contrary, there is every indication that their political elites and security structures have tacitly adopted a common approach to climate change.

But is it conceivable that any branch of government in an "open society" would covertly adopt a posture on a matter of such importance? That surely is not how liberal democracies are supposed to work?

Or have they ever really worked as they were supposed to? It is in the colonies, as Sartre once said, that the truths of the metropolis are most visible, and it is a fact certainly that the forms of statecraft that Britain used in its colonies were quite different from those of the metropole. This fissure was laid bare as far back as 1788, when Warren Hastings, the former governor of Bengal, was impeached by Edmund Burke on counts that amounted precisely to the charge that Hastings's statecraft in India represented an affront to the British political system. With Hastings's acquittal, the split came to be embedded at the heart of the imperial practices of the Anglosphere: through the nineteenth century and much of the twentieth, the statecraft that England and her settler colonies practiced in their dealings with non-Europeans was of an entirely different order from that which obtained domestically. Outside metropolitan areas, the functioning of power was always guided, in the first instance, by considerations of security. The maintenance of dominance outweighed any other imperative of governance, and it was toward these ends that statecraft was primarily oriented.

When seen through this prism, it does not seem at all improbable that certain organs of state, particularly the security establishment, would adopt an approach that is quite different from that of the domestic political sphere. Global warming is unique, after all, in that it is simultaneously a domestic and global crisis: a bifurcation of responses is only to be expected.

Nor is it conceivable that institutions of governance in any contemporary nation could be indifferent to global warming. For if it is the case that "biopolitics" is central to the mission of modern governments, as Michel Foucault argued, then climate change represents a crisis of unprecedented magnitude for their practices of governance: to ignore this challenge would run counter to the evolutionary path of the modern nation-state.

Moreover, the climate crisis holds the potential of drastically reordering the global distribution of power as well as wealth. This is because the nature of the carbon economy is such that power, no less than wealth, is largely dependent on the consumption of fossil fuels. The world's most powerful countries are also oil states, Timothy Mitchell notes, and "without the energy they derive from oil their current forms of political and economic life would not exist." Nor would they continue to occupy their present positions in the global ranking of power.

This being the case if the emissions of some countries were to be curbed while the emissions of others were allowed to rise, then this would lead inevitably to a redistribution of global power. It is certainly no coincidence that the increase in the consumption of fossil fuels in China and India has already brought about an enormous change in their international influence.

These realities cast a light of their own on the question of climate justice. That justice should be aspired to is widely agreed; it could hardly be otherwise since this ideal lies at the heart of all contemporary claims of political legitimacy. How such an end could be reached is also well known: an equitable regime of emissions could be created through any one of many strategies, such as "contraction and convergence," for instance, or "a per capita climate accord," or a fair apportioning of the world's remaining "climate budget." But the resulting equity would lead not just to a redistribution of wealth but also to a recalibration of global power—and from the point of view of a security establishment that is oriented toward the maintenance of global dominance, this is precisely the scenario that is most greatly to be feared; from this perspective the continuance of the status quo is the most desirable of outcomes.

Seen in this light, climate change is not a danger in itself; it is envisaged rather as a "threat multiplier" that will deepen already existing divisions and lead to the intensification of a range of conflicts. How will the security establishments of the West respond to these threat perceptions? In all likelihood they will resort to the strategy that Christian Parenti calls the "politics of the armed lifeboat," a posture that combines "preparations for open-ended counter-insurgency, militarized borders, [and] aggressive anti-immigrant policing." The tasks of the nation-state under these circumstances will be those of keeping "blood-dimmed tides" of climate refugees at bay and protecting their own resources: "In this world view, humanity has not only declared a war against itself, but is also locked into mortal combat with the earth."

The outlines of an "armed lifeboat" scenario can already be discerned in the response of the United States, United Kingdom, and Australia to the Syrian refugee crisis: they have accepted very few migrants even though the problem is partly of their own making. The adoption of this strategy might even represent the logical culmination of the biopolitical mission of the modern nation-state, since it is a strategy that conceives of the preservation of the "body of the nation" in the most literal sense: by a reinforcement of boundaries that are seen to be under threat from the infiltration of the pathological "bare life" that is spilling over from other nations.

The trouble, however, is that the contagion has already occurred, everywhere: the ongoing changes in the climate, and the perturbations that they will cause *within* nations, cannot be held at bay by reinforcing man-made boundaries. We are in an era when the body of the nation can no longer be conceived of as consisting only of a territorialized human population: its very sinews are now revealed to be intertwined with forces that cannot be confined by boundaries.

6.

It goes without saying that if the world's most powerful nations adopt the "politics of the armed lifeboat," explicitly or otherwise, then millions of people in Asia, Africa, and elsewhere will face doom. Unthinkable though this may appear, such a Darwinian approach would not be in conflict with free market ideology: that is why it has a long pedigree in the statecraft of the Anglosphere. Lest this seems far-fetched, let us recall that this is not the first time that British and American officialdom has had to confront catastrophes brought on by vagaries of climate. In the nineteenth and early twentieth centuries, El Niño events caused enormous disruption in India and the Philippines, and as Mike Davis has shown in his remarkable study *Late Victorian Holocausts*, in dealing with drought and famine, British and American colonial officials consistently placed far greater store on the sanctity of the free market than on human life. In these instances, as with the famines of Mao's China and Stalin's USSR, ideology prevailed over the preservation of life.

Malthusian ideas were also often invoked in the context of famine and starvation in Asia and Africa, as, for example, by Winston Churchill when he said, "Famine or no famine, Indians will breed like rabbits." Although we are unlikely to hear words of this kind in our era, there can be little doubt that there are many who believe that a Malthusian "correction" is the only hope for the continuance of "our way of life."

From this perspective, global inaction on climate change is by no means the result of confusion or denialism or a lack of planning: to the contrary, the maintenance of the status quo *is* the plan. Climate change may itself facilitate the realization of this plan by providing an alibi for ever-greater military intrusion into every kind of geographic and military space. And it is quite likely that this plan commands widespread but tacit support in many Western countries. Significant sections of the electorate probably understand that climate change negotiations may have the effect of changing their country's standing in the world's hierarchies of power as well as wealth: this may indeed form the basis of their resistance to climate science in general.

The refusal to acknowledge these realities sometimes lends an air of unreality to discussions of climate change. There are some who believe, for instance, that considerations of fairness may make people more willing to accept serious mitigatory measures. The trouble with this, in relation to climate justice, is that these measures would affect some far more than others. The geologist David Archer reckons that to reach a genuinely fair solution to the problem of emissions would "require cuts in the developed world of about 80 percent. For the United States, Canada and Australia, the cuts would be closer to 90 percent." Will an abstract idea of fairness be sufficient for people to undertake cuts on this scale, especially in a world where the pursuit of self-interest is conceived of as the motor of the economy? Let's just say there is much room for doubt.

The fact is that we live in a world that has been profoundly shaped by empire and its disparities. Differentials of power between and within nations are probably greater today than they have ever been. These differentials are, in turn, closely related to carbon emissions. The distribution of power in the world therefore lies at the core of the climate crisis. This is indeed one of the greatest obstacles to mitigatory action, and all the more so because it remains largely unacknowledged. This question will probably be even more difficult to resolve than economic disparities and matters like compensation, carbon budgets, and so on. We do at least possess a vocabulary for economic issues; within the current system of international relations, there is no language in which questions related to the equitable distribution of power can be openly and frankly addressed.

It is for these reasons that I differ with those who identify capitalism as the principal fault line on the landscape of climate change. It seems to me that this landscape is riven by two interconnected but equally important rifts, each of which follows a trajectory of its own: these are capitalism and empire (the latter being understood as an aspiration to dominance on the part of some of the most important structures of the world's most powerful states). In short, even if capitalism were to be magically transformed tomorrow, the imperatives of political and military dominance would remain a significant obstacle to progress on mitigatory action.

7.

The cynicism of the politics of the armed lifeboat is matched, on the other side, by the strategy that the elites of some large developing countries, like India, seem to be inclining toward: a politics of attrition. The assumption underlying this is that the populations of poor nations, because they are accustomed to hardship, possess the capacity to absorb, even if at great cost, certain shocks and stresses that might cripple rich nations.

This may not be as delusional as it sounds. It is not impossible, for instance, that in dealing with situations of extraordinary stress the very factors that are considered advantages in coping with extreme weather—education, wealth, and a high degree of social organization—may actually become vulnerabilities. Western food production, for instance, is dangerously resource intensive, requiring something in the range of a "dozen fossil fuel calories for each food calorie." And Western food distribution systems are so complex that small breakdowns could lead to cascading consequences that culminate in complete collapse. Power failures, for instance, are so rare in advanced countries that they often cause great disruption—including spikes in rates of crime—when they do occur. In many parts of the global south, breakdowns are a way of life, and everybody is used to improvisations and work-arounds.

In poor countries, even the middle classes are accustomed to coping with shortages and discomforts of all sorts; in the West, wealth, and habits based upon efficient infrastructures, may have narrowed the threshold of bearable pain to a point where climatic impacts could quickly lead to systemic stress.

Acclimatization to difficult conditions may itself produce certain sorts of resilience, especially in regard to one of the most immediate effects of global warming: extreme heat. Thus, for instance, the European heat wave of 2003 resulted in forty-six thousand deaths, while the 2010 heat wave in Russia had an estimated death toll of fifty-six thousand. These figures are far in excess of the toll of the 2015 heatwave in South Asia and the Persian Gulf region which registered heat index readings of as much as 163 degrees Fahrenheit (72.8 degrees Celsius). Moreover, ties of community are still strong through the global south; people who are completely cut off from others are relatively rare. This too is a safety net of a kind: recent experience shows that the absence of community networks can greatly amplify the impact of extreme weather events. After the 2003 heat wave in Europe, for instance, it was found that many of the dead were elderly people living in isolation.

In short: the rich have much to lose; the poor do not. This is true not just of international relations but also of the internal structure of the developing world, where the urban middle classes have a carbon footprint that is not much lower than that of the average European. However, it is not the middle classes and the political elites of the global south that will bear the brunt of the suffering but rather the poor and the disempowered. This too is a brake on effective progress in climate negotiations in that it reduces the incentive to compromise: the belief that they are not gambling with their own lives is, no doubt, just as important a factor for the political elites of the developing world as it is for their counterparts in the West. It is therefore not totally unrealistic to assume that poor countries may be able to force rich countries to make greater concessions merely by absorbing the impacts of climate change, at no matter what cost.

These considerations are, as I have noted, just as cynical as those that underlie the politics of the armed lifeboat. Yet, it is hard also to determine what an ethical strategy might be for poor countries like India. Should they perhaps abandon the quest for Western-style prosperity, so that a greater number will survive to take the struggle for justice forward in some uncertain future? But this would require the abandonment also of the project of "modernization" that was often implicit in decolonization: it would put a freeze on a system of colonial-style inequality.

In any case, who could possibly make a convincing case for the poor to make sacrifices so that the rich can continue to enjoy the fruits of their wealth? To do so would be an acknowledgment that the ideas of equality and justice from which the dominant political imaginary draws its legitimacy have never been anything other than grotesque fictions, designed to secure exactly the opposite of those professed ends. This perhaps is why such a case is never explicitly made but only implied by euphemistic exhortations that urge poor countries to take a "different road to development" and so on.

Take the use of coal. Much concern has been expressed in the West about coal plants in India. Yet, analysts have calculated that "in 2014 the average Indian accounted for around 20 per cent of the average American's coal consumption and around 34 per cent of those from the OECD." The logical and equitable response might be for the United States or the Organization for Economic Co-operation and Development to shut down one of their coal plants every time a new plant is commissioned in India, until a convergence occurs. But this is, of course, highly unlikely to happen.

This then is another way in which the terrain of global warming has been shaped, not just by capitalism but also by empire: the impetus for industrialization in much of the world was a part of the trajectory of decolonization, and the historical legacy of those conflicts is also embedded in the context of climate change negotiations. The end result is that these negotiations now resemble a form of high-stakes gambling in which catastrophe is the card that is expected to trump all others.

8.

In the annals of climate change, 2015 was a momentous year. Extreme weather events abounded: a strong El Niño, perching upon "the ramp of global warming," wrought havoc upon the planet; many millions of people found themselves at the mercy of devastating floods and droughts; freakish tornadoes and cyclones churned through places where they had never been seen before; and extraordinary temperature anomalies were recorded around the globe, including unheard-of midwinter highs over the North Pole. Within days of the year's end, 2015 was declared the hottest year since record-keeping began. It was a year in which the grim predictions of climate scientists assumed the ring of prophecy.

These disturbances were almost impossible to ignore: on the web as in the traditional media the phrase "climate change" was everywhere. Few indeed were the quarters that remained unperturbed, but literary fiction and the arts appear to have been among them: short lists for prizes, reviews, and so on, betray no signs of a heightened engagement with climate change.

But 2015 did produce two very important publications on climate change: the first, Pope Francis's encyclical letter *Laudato Si'*, was published in May; while the second, the Paris Agreement on climate change, appeared in December.

These two documents occupy a realm that few texts can aspire to: one in which words effect changes in the real world. But the documents are also *texts*, brought into being through the crafts of writing, with meticulous attention being paid to form, vocabulary, and even typography. To read them as texts is revealing in many ways.

As is only to be expected, the two works, one written by a former teacher of literature and the other by a multitude of diplomats and delegates, are not at all similar, even though they rely on many of the same materials and address some of the same subjects. Yet they also have certain things in common: perhaps the most important of these is that they are both founded on an acceptance of the research produced by climate science. In this sense they together represent a historic milestone: their publication marks a general, worldwide acknowledgment that the earth's climate is changing and that human beings are largely responsible for these changes. The documents can therefore rightly be seen as a vindication of climate science.

Beyond that, the documents diverge sharply, although not in predictable ways. It might be thought, for example, that as a primarily religious document the pope's Encyclical would be written in an allusive and ornate style; it might equally be expected that the Agreement would, by contrast, be terse and workmanlike (as was the Kyoto Protocol, for instance). In fact the opposite is true. The Encyclical is remarkable for the lucidity of its language and the simplicity of its construction; it is the Agreement, rather, that is highly stylized in its wording and complex in structure.

The Agreement is divided into two parts: the first and longer part is entitled "Proposal by the President," while the second—which is the Agreement itself—is described as an "Annex." Each part is preceded by a preamble, as is the convention for treaties—except that in this case these sections are far longer and more elaborate than is customary. The preamble to the Kyoto Protocol, for instance, consists of only five terse declarative clauses; by contrast, the text of the Paris Agreement contains no less than thirty-one ringing declarations. Fifteen of these precede the first part of the document (the president's proposal); here are some of them:

Recalling decision 1/CP.17 on the establishment . . .
Also recalling Articles . . .
Further recalling relevant decisions . . .
Welcoming the adoption . . .
Recognizing that . . .
Acknowledging that . . .
Agreeing to uphold and promote . . .

The lines pour down the page in a waterfall of gerunds and then, without the sentence yet reaching an end, the clauses change into numbered articles as the document switches gear and "*Decides* to adopt . . ." and "*Requests* the Secretary-General . . ."

And so the Proposal continues, covering eighteen densely printed pages: yet this large block of text, with its 140 numbered clauses and six sections, consists of only two sentences, one of which runs on for no less than fifteen pages! Indeed this part of the Agreement is a work of extraordinary compositional virtuosity—thousands of words separated by innumerable colons, semicolons, and commas and only a single, lonely pair of full stops.

The giddy virtuosity of the text provides a context for the images that streamed out of Paris after the negotiations: of world leaders and business tycoons embracing each other; of negotiators with tears in their eyes; of delegates crowding joyfully together to be photographed. The pictures captured a mood of as much astonishment as joy; it was as if the delegates could not quite believe that they had succeeded in reaching an agreement of such significance. The euphoria that resulted is as clearly evident in the text of the Agreement as it is in the pictures: the virtuosity of its composition is a celebration of its own birth.

There is no such exuberance in *Laudato Si'*, which is remarkable instead for the sober clarity with which it addresses complex questions. While the preambles of the Agreement occupy a prosodic domain of their own, somewhere between poetry and prose, *Laudato Si'* resorts to poetry only at the very end, in two concluding prayers.

Here again lies an unexpected difference between the two documents. Because of the prayerful ending of *Laudato Si'*, it might be thought that there would be more wishful thinking and conjecture in the Encyclical than in the Agreement. But that too is by no means the case. It is the Paris Agreement rather that repeatedly invokes the impossible: for example, the aspirational goal of limiting the rise in global mean temperatures to 1.5 degrees Centigrade—a target that is widely believed to be already beyond reach.

Although the Paris Agreement does not lay out the premises on which its targets are based, it is thought that they are founded on the belief that technological advances will soon make it possible to whisk greenhouse gases out of the atmosphere and bury them deep underground. But these technologies are still in their nascency, and the most promising of them, known as "biomass energy carbon capture and storage," would require the planting of bioenergy crops over an area larger than India to succeed at scale. To invest so much trust in what is yet only a remote possibility is little less than an act of faith, not unlike religious belief.

Laudato Si', by contrast, does not anywhere suggest that miraculous interventions may provide a solution for climate change. It strives instead to make sense of humanity's present predicament by mining the wisdom of a tradition that far predates the carbon economy. Yet it does not hesitate to take issue with past positions of the Church, as, for example, in the matter of reconciling an ecological consciousness with the Christian doctrine of Man's dominion over Nature. Even less does the

Encyclical hesitate to criticize the prevalent paradigms of our era; most of all it is fiercely critical of "the idea of infinite or unlimited growth, which proves so attractive to economists, financiers and experts in technology." It returns to this theme repeatedly, insisting that it is because of the "technocratic paradigm" that "we fail to see the deepest roots of our present failures, which have to do with the direction, goals, meaning and social implications of technological growth."

In the text of the Paris Agreement, by contrast, there is not the slightest acknowledgment that something has gone wrong with our dominant paradigms; it contains no clause or article that could be interpreted as a critique of the practices that are known to have created the situation that the Agreement seeks to address. The current paradigm of perpetual growth is enshrined at the core of the text.

But perhaps criticism is not the business of a treaty? Not true: international narcotics agreements, for example, use quite strong language in condemning "the evil of drug addiction," and so on. Critical language even figured in earlier climate treaties like the Kyoto Protocol, which did make reference to "market imperfections." No such phrase is to be found in the Paris Agreement: it merely acknowledges that "climate change is a common concern for humankind."

The Agreement is similarly tepid in its naming of the conditions that it is intended to remedy: while words like *catastrophe* and *disaster* occur several times in the Encyclical, the Agreement speaks only of the *adverse impacts* or *effects* of climate change. The word *catastrophe* is never used and even *disaster* occurs only once, and that too only because it figures in the title of a previous conference. It is as if the negotiations had been convened to deal with a minor annoyance. No wonder then that the Agreement's provisions will come into force (if such a word can be used of voluntary actions) only in 2020 when the window for effective action will already have dwindled to the size of the eye of a celestial needle.

In contrast to the Agreement's careful avoidance of disruptive terminology, *Laudato Si'* challenges contemporary practices not just in its choice of words but also in the directness of its style. In place of the obscurity and technical jargon that enshrouds the official discourse on climate change, the document strives to open itself, in a manner that explicitly acknowledges the influence of the saint who is the pope's "guide and inspiration": "Francis [of Assisi] helps us to see that an integral ecology calls for openness to categories which transcend the language of mathematics and biology and take us to the heart of what it is to be human."

In much the same measure that *Laudato Si'* strives for openness, the Agreement moves in the opposite direction: toward confinement and occlusion. Its style as well as its vocabulary convey the impression of language being deployed as an instrument of concealment and withdrawal; even its euphoria is suggestive of the heady joy of a small circle of initiates celebrating a rite of passage. In clause after clause, the Agreement summons up mysterious structures, mechanisms, and strange new avatars of officialdom—as, for example, when it "*decides* that two high-level champions shall be appointed," and "*invites* all interested parties . . . to support the work of the champions" (where, one wonders, is the Colosseum in which these champions have dueled their way to the "highest level"?).

That the word *champion* is left undefined is telling: it implies that the document's authors know tacitly whom they are referring to—and who could that be but others like themselves? This is indeed an Agreement of champions, authored by and for those of that ilk.

Strangely, *Laudato Si'* seems to anticipate this possibility: in a passage that refers to the way that decisions are made in "in-

ternational political and economic discussions," it points to the
role of "professionals, opinion makers, communications media
and centres of power [who] being located in affluent urban areas, are far removed from the poor, with little direct contact
with their problems. They live and reason from the comfortable position of a high level of development and a quality of
life well beyond the reach of the majority of the world's population." It is with exactly this in mind that the style of *Laudato
Si'* seems to have been forged, as an attempt to address those
to whom it repeatedly refers as the "excluded."

The opacity of the Agreement, on the other hand, hints at
the opposite intention: its rhetoric is like a shimmering screen,
set up to conceal implicit bargains, unspoken agreements, and
loopholes visible only to those in the know. It is no secret that
various billionaires, corporations, and "climate entrepreneurs"
played an important part in the Paris negotiations. But even if
this were not publicly known, it would be deducible from the
diction of the Agreement, which is borrowed directly from the
free-trade agreements of the neo-liberal era: these clearly are
the provenance of its references to "accelerating, encouraging
and enabling innovation" and of many of the terms on which
it relies, such as *stakeholder, good practices, insurance solutions,
public and private participation, technology development,* and so on.

As is often the case with texts, the Agreement's rhetoric
serves to clarify much that it leaves unsaid: namely, that its
intention, and the essence of what it has achieved, is to create
yet another neo-liberal frontier where corporations, entrepreneurs, and public officials will be able to join forces in enrich-
ing each other.

Might the Paris Agreement have taken a different turn if the
terrorist attacks of December 2 had not radically changed the
context of the negotiations by providing the French government with an alibi for the banning of demonstrations, marches,

and protests? What would have happened if the delegates had been forced to deal with a great wave of popular pressure, as climate activists had planned? These questions will haunt historians for years to come, and the answers, of course, will never be known. However, the alacrity with which the French authorities moved against climate activists, and the efficiency with which it put dozens of them under house arrest, suggests that even in the absence of the attacks a means would have been found for corralling the protesters—as has been the case at many other international negotiations during the last two decades. This is one area in which governments and corporations around the world have grown extraordinarily skilled, and there is every reason to believe that the investments that they have made in surveilling environmental activists would have paid off, once again, to enforce the exclusions that are hinted at in the Agreement's text.

If exclusion is a recurrent theme in *Laudato Si'*, it is for exactly the opposite reason: because poverty and justice are among the Encyclical's central concerns. The document returns over and again to the theme of "how inseparable the bond is between concern for nature, justice for the poor, commitment to society, and interior peace."

In *Laudato Si'* the words *poverty* and *justice* keep close company with each other. Here poverty is not envisaged as a state that can be managed or ameliorated in isolation from other factors; nor are ecological issues seen as problems that can be solved without taking social inequities into account, as is often implied by a certain kind of conservationism. *Laudato Si'* excoriates this latter kind of "green rhetoric" and insists that "a true ecological approach *always* becomes a social approach; it must integrate questions of justice in debates on the environment, so as to hear *both the cry of the earth and the cry of the poor.*" This in turn leads to the blunt assertion that "a true 'ecological debt' exists, particularly between the global north and south."

Here again the contrast with the Paris Agreement is stark. When poverty finds mention in the Agreement, it is always as a state in itself, to be alleviated through financial and other mechanisms. The word never occurs in connection with *justice*—but this is scarcely surprising since there is only one mention of justice in the text and that too in a clause that is striking for the care with which it is worded: the preamble to the Annex merely takes note of "the importance for some of the concept of 'climate justice' when taking action to address climate change."

The scare quotes that bracket the phrase "climate justice" and the description of the concept as being important only "for some" amount to nothing less than an explicit disavowal of the concept. But an implicit disavowal occurs much earlier, in one of the few passages in the text that is pellucid in its clarity: "the Agreement does not involve or provide a basis for any liability or compensation." With these words the Agreement forever strips the victims of climate change of all possible claims to legal recompense for their losses; they will have to depend instead on the charity of a fund that developed nations have agreed to set up.

The differences between the two texts is never clearer than in the manner of their endings. The Agreement concludes by conjuring itself into being through the will of the signatories and by announcing the date of its self-actualization: the twelfth day of December, in the year 2015. The very syntax is an expression of faith in the sovereignty of Man and his ability to shape the future.

The prayers with which *Laudato Si'* concludes, on the other hand, are an appeal for help and guidance. As such they are also acknowledgments of how profoundly humanity has lost its way and of the limits that circumscribe human agency. In this they echo one of the most radical elements of Pope Francis's critique of the era that he describes as "a period of irrational confidence in progress and human abilities." It is his questioning of the idea that "human freedom is limitless." "We have forgotten," goes the text, "that 'man is not only a freedom which he creates for himself. . . . He is spirit and will, but also nature.'"

It is by this route that the themes of *Laudato Si'* lead back to the territory that I explored earlier in trying to locate the fronts where climate change resists contemporary literature and the arts. Insofar as the idea of the limitlessness of human freedom is central to the arts of our time, this is also where the Anthropocene will most intransigently resist them.

9.

Bleak though the terrain of climate change may be, there are a few features in it that stand out in relief as signs of hope: a spreading sense of urgency among governments and the public; the emergence of realistic alternative energy solutions; widening activism around the world; and even a few signal victories for environmental movements. But the most promising development, in my view, is the increasing involvement of religious groups and leaders in the politics of climate change. Pope Francis is, of course, the most prominent example, but some Hindu, Muslim, Buddhist, and other groups and organizations have also recently voiced their concern.

I take this to be a sign of hope because it is increasingly clear to me that the formal political structures of our time are incapable of confronting this crisis on their own. The reason for this is simple: the basic building block of these structures is the nation-state, inherent to the nature of which is the pursuit of the interests of a particular group of people. So powerful is this imperative that even transnational groupings of nation-states, like the UN, seem unable to overcome it. This is partly due, of course, to questions of power and geo-political rivalries. But it may also be that climate change represents, in its very nature, an unresolvable problem for modern nations in terms of their biopolitical mission and the practices of governance that are associated with it.

I would like to believe that a great upsurge of secular protest movements around the world could break through the deadlock and bring about fundamental changes. The problem, however, is time. One of the reasons why climate change is a "wicked" as opposed to a "normal" problem is that the time horizon in which effective action can be taken is very narrow: every year that passes without a drastic reduction in global emissions makes catastrophe more certain.

It is hard to see how popular protest movements could gain enough momentum within such a narrow horizon of time: such movements usually take years, even decades, to build. And to build them in the current situation will be all the more difficult because security establishments around the world have already made extensive preparations for dealing with activism.

If a significant breakthrough is to be achieved, if the securitization and corporatization of climate change is to be prevented, then already-existing communities and mass organizations will have to be in the forefront of the struggle. And of such organizations, those with religious affiliations possess the ability to mobilize people in far greater numbers than any others. Moreover, religious worldviews are not subject to the limitations that have made climate change such a challenge for our existing institutions of governance: they transcend nation-states, and they all acknowledge intergenerational, long-term responsibilities; they do not partake of economistic ways of thinking and are therefore capable of imagining nonlinear change—catastrophe, in other words—in ways that are perhaps closed to the forms of reason deployed by contemporary nation-states. Finally, it is impossible to see any way out of this crisis without an acceptance of limits and limitations, and this in turn, is, I think, intimately related to the idea of the sacred, however one may wish to conceive of it.

If religious groupings around the world can join hands with popular movements, they may well be able to provide the momentum that is needed for the world to move forward on drastically reducing emissions without sacrificing considerations of equity. That many climate activists are already proceeding in this direction is, to me, yet another sign of hope.

The ever-shrinking time horizon of the climate crisis may itself be a source of hope in at least one sense. Over the last few decades, the arc of the Great Acceleration has been completely in line with the trajectory of modernity: it has led to the destruction of communities, to ever greater individualization and anomie, and to the industrialization of agriculture and to the centralization of distribution systems. At the same time, it has also reinforced the mind-body dualism to the point of producing the illusion, so powerfully propagated in cyberspace, that human beings have freed themselves from their material circumstances to the point where they have become floating personalities "decoupled from a body." The cumulative effect is the extinction of exactly those forms of traditional knowledge, material skills, art, and ties of community that might provide succor to vast numbers of people around the world— and especially to those who are still bound to the land—as the impacts intensify. The very speed with which the crisis is now unfolding may be the one factor that will preserve some of these resources.

The struggle for action will no doubt be difficult and hard-fought, and no matter what it achieves, it is already too late to avoid some serious disruptions of the global climate. But

I would like to believe that out of this struggle will be born a generation that will be able to look upon the world with clearer eyes than those that preceded it; that they will be able to transcend the isolation in which humanity was entrapped in the time of its derangement; that they will rediscover their kinship with other beings, and that this vision, at once new and ancient, will find expression in a transformed and renewed art and literature.

ACKNOWLEDGMENTS

This book began as a set of four lectures, presented at the University of Chicago in the fall of 2015. The lectures were the second in a series named after the family of Randy L. and Melvin R. Berlin. I am deeply grateful to the Berlin family, and the administrators of the series, for providing me with an opportunity to develop my ideas on climate change.

There could have been no more congenial milieu in which to present these ideas than the University of Chicago, which is a global pioneer in the study of the Anthropocene. The comments and questions of those who attended the lectures gave me a great deal to think about, as did my interactions with the university community. My thanks go especially to Dipesh Chakrabarty, Julia Adeney Thomas, and Kenneth Pomeranz: they have all enriched this book, through their work in the first instance, but also by their comments on the lectures.

Prasenjit Duara and Tansen Sen gave me invaluable guidance and advice on Chinese materials; they and Liang Yongjia also translated certain passages that are quoted in the book. I owe them many thanks.

Through much of the time that I was working on and editing this book, I was a Visiting Fellow at the Ford Foundation: I gratefully acknowledge the foundation's support.

My wife, Deborah Baker, read an early version of the manuscript, as did Adam Sobel, Rahul Srivastava, and Mukul Kesavan; Lucano Alvares and Raghu Kesavan pointed me to some important sources: I owe them all many thanks.

Studies have shown that a mention of global warming at a dinner table is almost certain to lead to a quick change of subject. My children, Lila and Nayan, did not always have that option when they were growing up: I am, as ever, grateful to them for their forbearance.

My thanks go finally to Alan Thomas and Meru Gokhale, my editors, and to the three anonymous readers who reviewed the manuscript for the University of Chicago Press: their comments were invaluable.

NOTES

PART I

4 ignorance to knowledge: "Recognition . . . is a change from
ignorance to knowledge, disclosing either a close relationship or
enmity, on the part of people marked out for good or bad fortune."
Aristotle, *Poetics*, tr. Malcolm Heath (London: Penguin 1996), 18.

5 lies within oneself: In the phrasing of Giorgio Agamben, the
philosopher, these are moments in which potentiality turns "back
upon itself to give itself to itself" (*Homo Sacer: Sovereign Power
and Bare Life*, tr. Daniel Heller-Roazen [Stanford, CA: Stanford
University Press, 1998], 46).

7 genre of science fiction: Barbara Kingsolver's *Flight Behavior* and
Ian McEwan's *Solar*, both of which were widely reviewed by literary
journals, are rare exceptions.

7 feedback loop: In Gavin Schmidt and Joshua Wolfe's definition:
"The concept of feedback is at the heart of the climate system and
is responsible for much of its complexity. In the climate everything
is connected to everything else, so when one factor changes, it
leads to a long chain of changes in other components, which
leads to more changes, and so on. Eventually, these changes end
up affecting the factor that instigated the initial change. If this
feedback amplifies the initial change, it's described as positive,
and if it dampens the change, it is negative." See *Climate Change:
Picturing the Science*, ed. Gavin Schmidt and Joshua Wolfe (New York:
W. W. Norton, 2008), 11.

8 wild has become the norm: Lester R. Brown writes, "climate
instability is becoming the new norm." See *World on the Edge: How*

to Prevent Environmental and Economic Collapse (New York: W. W. Norton, 2011), 47.

8 "stories our civilization tells itself": See dark-mountain.net; and see also John H. Richardson, "When the End of Human Civilization Is Your Day Job," *Esquire*, July 7, 2015.

9 era of the Anthropocene: Dipesh Chakrabarty, "The Climate of History: Four Theses," *Critical Inquiry* 35 (Winter 2009).

9 "processes of the earth": The quote is from Naomi Oreskes, "The Scientific Consensus on Climate Change: How Do We Know We're Not Wrong," in *Climate Change: What It Means for Us, Our Children and Our Grandchildren*, ed. Joseph F. C. DiMento and Pamela Doughman (Cambridge, MA: MIT Press, 2007). For a discussion of the genealogy of the concept of the Anthropocene, see Paul J. Crutzen, "Geology of Mankind," *Nature* 415 (January 2002): 23; and Will Steffen, Jacques Grinevald, Paul Crutzen, and John McNeill, "The Anthropocene: Conceptual and Historical Perspectives," *Philosophical Transactions of the Royal Society* 369 (2011): 842–67.

10 the wind in our hair: Stephanie LeMenager calls this "the road-pleasure complex" in *Living Oil: Petroleum Culture in the American Century* (Oxford: Oxford University Press, 2014,) 81.

11 Bangkok uninhabitable: Cf. James Hansen: "'Parts of [our coastal cities] would still be sticking above the water, but you couldn't live there." http://www.thedailybeast.com/articles/2015/07/20/climate-seer-james-hansen-issues-his-direst-forecast-yet.html.

11 Great Derangement: As the historian Fredrik Albritton Jonsson notes, if we consider the transgression of the "planetary boundaries that are necessary to maintain the Earth system 'in a Holocene-like state' . . . our current age of fossil fuel abundance resembles nothing so much as a giddy binge rather than a permanent achievement of human ingenuity" ("The Origins of Cornucopianism: A Preliminary Genealogy," *Critical Historical Studies*, Spring 2014, 151).

14 meteorological history: The only part of the Indian subcontinent where tornadoes occur frequently is in the Bengal Delta, particularly Bangladesh. Cf. Someshwar Das, U. C. Mohanty, Ajit Tyagi, et al., "The SAARC Storm: A Coordinated Field Experiment

on Severe Thunderstorm Observations and Regional Modeling over the South Asian Region," *American Meteorological Society*, April 2014, 606.

16 "being aware of it": Ian Hacking, *The Emergence of Probability* (Cambridge: Cambridge University Press, 1975), Kindle edition, loc. 194.

17 "into the foreground": Franco Moretti, "Serious Century: From Vermeer to Austen," in *The Novel, Volume 1*, ed. Franco Moretti (Princeton, NJ: Princeton University Press, 2006), 372.

17 regime of thought and practice: Cf. Giorgio Agamben on Carl Schmitt, "the true life of the rule is the exception," in *Homo Sacer*, tr. Daniel Heller-Roazen, 137.

18 "pictures of Bengali life": Bankim Chandra Chatterjee, "Bengali Literature," first published anonymously in *Calcutta Review* 104 (1871). Digital Library of India: http://en.wikisource.org/wiki/ Bengali_Literature.

18 early 1860s: See also my essay, "The March of the Novel through History: The Testimony of My Grandfather's Bookcase," in the collection *The Imam and the Indian* (New Delhi: Ravi Dayal, 2002).

19 "blond cornfields": Gustave Flaubert, *Madame Bovary*, tr. Eleanor Marx-Aveling (London: Wordsworth Classics, 1993), 53.

19 "no miracles at all": Franco Moretti, *The Bourgeois* (London: Verso, 2013), 381. There is an echo here of Carl Schmitt's observation: "The idea of the modern constitutional state triumphed together with deism, a theology and metaphysics that banished the miracle from the world. . . . The rationalism of the Enlightenment rejected the exception in every form" (*Political Theology: Four Chapters on the Concept of Sovereignty* [University of Chicago Press, 2005], 36–37).

20 "change in the present": Spencer R. Weart, *The Discovery of Global Warming* (Cambridge, MA: Harvard University Press, 2003), 9.

20 "does not make leaps": Stephen Jay Gould, *Time's Arrow, Time's Cycle: Myth and Metaphor in the Discovery of Geological Time* (Cambridge, MA: Harvard University Press, 1987), 173.

20 jump, if not leap: The theory of punctuated equilibrium, as articulated by Stephen Jay Gould and Niles Eldredge, proposed "that the emergence of new species was not a constant process but

moved in fits and starts: it was not gradual but punctuated." See
John L. Brooke, *Climate Change and the Course of Global History: A
Rough Journey* (New York: Cambridge University Press, 2014), 29.

20   "'both and neither'": Gould, *Time's Arrow, Time's Cycle*, 191.

20   "short-lived cataclysmic events": http://geography.about.com/od/
physicalgeography/a/uniformitarian.htm.

21   "immaterial and supernatural agents": Gould, *Time's Arrow, Time's
Cycle*, 108–9.

21   "victim with her cold beams": Chatterjee, "Bengali Literature."

21   "nightingales in shady groves": Flaubert, *Madame Bovary*, 28.

22   "resent its interference": Chatterjee, "Bengali Literature."

22   "reigned supreme": Brooke, *Climate Change and the Course of Global
History*.

22   "events in the stars": Quoted in Gould, *Time's Arrow, Time's Cycle*, 176.

23   "covers of popular magazines": Elizabeth Kolbert, *The Sixth
Extinction: An Unnatural History* (New York: Henry Holt, 2014), 76. See
also Jan Zalasiewicz and Mark Williams, *The Goldilocks Planet: The
Four Billion Year Story of Earth's Climate* (Oxford: Oxford University
Press, 2012), Kindle edition, loc. 3042, and Gwynne Dyer, *Climate
Wars: The Fight for Survival as the World Overheats* (Oxford: Oneworld
Books, 2010), Kindle edition, loc. 3902.

23   "basis of intelligibility": Gould, *Time's Arrow, Time's Cycle*, 10.

23   "'carry me with you!'": Flaubert, *Madame Bovary*, 172–73.

25   recorded meteorological history: Adam Sobel, *Storm Surge: Hurricane
Sandy, Our Changing Climate, and Extreme Weather of the Past and
Future* (New York: HarperCollins, 2014), Kindle edition, locs. 91–105.

25   its impacts: Ibid., locs. 120, 617–21.

26   "faraway places": Ibid., loc. 105.

26   "possible in Brazil": Mark Lynas, *Six Degrees: Our Future on a Hotter
Planet* (New York: HarperCollins, 2008), 41.

26   named the "catastophozoic": Kolbert, *The Sixth Extinction*, 107.

26   "the long emergency" and "Penumbral Period": David Orr, *Down
to the Wire: Confronting Climate Collapse* (Oxford: Oxford University
Press, 2009), 27–32; and Naomi Oreskes and Erik M. Conway, *The
Collapse of Western Civilization: A View from the Future* (New York:
Columbia University Press, 2014), 4.

26 "extremes of heat and cold": Geoffrey Parker, *Global Crisis: War, Climate Change, and Catastrophe in the Seventeenth Century* (New Haven, CT: Yale University Press, 2013), Kindle edition, loc. 17574.

28 were killed by tigers: In his book, *The Royal Tiger of Bengal: His Life and Death* (London: J. and A. Churchill, 1875), Joseph Fayrer records that between 1860 and 1866 4,218 people were killed by tigers in Lower Bengal.

29 this fearsome sight: Amitav Ghosh, *The Hungry Tide* (New York: Houghton Mifflin Harcourt, 2005).

30 "feels it generally": Martin Heidegger, *Existence and Being*, intro. Werner Brock, tr. R. F. C. Hull and Alan Crick (Washington, DC: Gateway Editions, 1949), 336.

30 "something uncanny": Timothy Morton, *Hyperobjects* (Minneapolis: University of Minnesota Press, 2013), Kindle edition, loc. 554.

30 "menace and uncertainty": George Marshall, *Don't Even Think about It: Why Our Brains Are Wired to Ignore Climate Change* (New York: Bloomsbury, 2014), 95.

31 processes of thought: Cf. Eduardo Kohn, *How Forests Think: Toward an Anthropology beyond the Human* (Berkeley: University of California Press, 2013).

33 relationship with the nonhuman: Cf. Michael Shellenberger and Ted Nordhaus, "The Death of Environmentalism: Global Warming Politics in a Post-Environmental World" (Oakland, CA: Breakthrough Institute, 2007): "The concepts of 'nature' and 'environment' have been thoroughly deconstructed. Yet they retain their mythic and debilitating power within the environmental movement and the public at large" (12).

33 "post-natural world": Bill McKibben, *The End of Nature* (New York: Random House, 1989), 49.

38 tides and the seasons: Anuradha Mathur and Dilip da Cunha make this point at some length in their excellent book, *SOAK: Mumbai in an Estuary* (New Delhi: Rupa Publications, 2009).

38 and on Salsette: I am grateful to Rahul Srivastava, the urban theorist and cofounder of URBZ (http://urbz.net/about/people/), for this insight.

38 a chest of tea: Bennett Alan Weinberg and Bonnie K. Bealer, *The*

World of Caffeine: The Science and Culture of the World's Most Popular Drug (New York: Routledge, 2000), 161.

39 "milieu of colonial power": Anuradha Mathur and Dilip da Cunha, *SOAK*, 47.

39 their colonial origins: The British geographer James Duncan describes the colonial city as a "political tract written in space and carved in stone. The landscape was part of the practice of power." Quoted in Karen Piper, *The Price of Thirst: Global Water Inequality and the Coming Chaos* (Minneapolis: University of Minnesota Press, 2014), Kindle edition, loc. 3168.

39 "an island once": Govind Narayan, *Govind Narayan's Mumbai: An Urban Biography from 1863*, tr. Murali Ranganathan (London: Anthem Press, 2009), 256. I am grateful to Murali Ranganathan for clarifying many issues relating to the topography of Mumbai.

39 "concentration of risk": Cf. Aromar Revi, "Lessons from the Deluge," *Economic and Political Weekly* 40, no. 36 (September 3–8, 2005): 3911–16, 3912.

40 cyclonic activity: A 2010 report published jointly by the India Meteorological Department and National Disaster Management Authority places the coastal districts of the India's western states in the lowest category of proneness to cyclones (table 9).

40 west coast of India: Earthquakes of 5.8 and 5.0 magnitude were recorded in the Owen fracture zone on October 2, 2013, and November 12, 2014. For details, see http://dynamic.pdc.org/snc/prod/40358/rr.html & http://www.emsc-csem.org/Earthquake/earthquake.php?id=408320.

40 "NW Indian Ocean": M. Fournier, N. Chamot-Rooke, M. Rodriguez, et al., "Owen Fracture Zone: The Arabia-India Plate Boundary Unveiled," *Earth and Planetary Science Letters* 302 nos. 1–2 (February 1, 2011): 247–52.

41 after the monsoons: Hiroyuki Murakami et al., "Future Changes in Tropical Cyclone Activity in the North Indian Ocean Projected by High Resolution MRI-AGCMs," *Climate Dynamics* 40 (2013): 1949–68, 1949.

41 region's wind patterns: Amato T. Evan, James P. Kossin, et al., "Arabian Sea Tropical Cyclones Intensified by Emissions of Black

Carbon and Aerosols," *Nature* 479 (2011): 94–98.

43 "minor cyclonic storms": *Gazetteer of Bombay City and Island,* Vol. I (1909), 96. I am grateful to Murali Ranganathan for providing me with this reference.

43 "end of all things": Quoted in *Gazetteer of Bombay City and Island,* Vol. I (1909), 97.

43 "persons were killed": Ibid., 98.

44 people were killed: Ibid., 99.

44 "number and intensity": Ibid.

44 intensity scale: On the Saffir-Simpson hurricane intensity scale, wind speeds of 75 mph are the benchmark for a Category 1 hurricane. In the Tropical Cyclone Intensity Scale used by the India Meteorological Department, any storm with wind speeds of over 39 kmph counts as a "cyclonic storm," hence this storm was named Cyclone Phyan.

45 single day: R. B. Bhagat et al., "Mumbai after 26/7 Deluge: Issues and Concerns in Urban Planning," *Population and Environment* 27, no. 4 (March 2006): 337–49, 340.

45 estuarine location: I am deeply grateful to Rahul Srivastava, Manasvini Hariharan, Apoorva Tadepalli, and the team at URBZ for their help with the research for this section.

45 filth-clogged ditches: In "Drainage Problems of Brihan Mumbai," B. Arunachalam provides a concise account of how Mumbai's hydrological systems have been altered over time (*Economic and Political Weekly* 40, no. 36 [September 3–9, 2005]: 3909–11, 3909).

45 absorptive ability: Cf. Vidyadhar Date, "Mumbai Floods: The Blame Game Begins," *Economic and Political Weekly* 40, no. 34 (August 20–26, 2005): 3714–16, 3716; and Ranger et al., "An Assessment of the Potential Impact of Climate Change on Flood Risk in Mumbai," *Climate Change* 104 (2011): 139–67, 142, 146; see also R. B. Bhagat et al., "Mumbai after 26/7 Deluge," 342.

46 1.5 million: P. C. Sehgal and Teki Surayya, "Innovative Strategic Management: The Case of Mumbai Suburban Railway System," *Vikalpa* 36, no. 1 (January–March 2011): 63.

46 knocked out as well: Aromar Revi, "Lessons from the Deluge," 3913.

46 suffered damage: The paragraphs above are based largely on the

Fact Finding Committee on Mumbai Floods, Final Report, vol. 1, 2006, 13–15.

47 fishing boat: Vidyadhar Date, "Mumbai Floods," 3714.

47 trapped by floodwaters: Aromar Revi, "Lessons from the Deluge," 3913.

47 homes to strangers: Cf. Carsten Butsch et al., "Risk Governance in the Megacity Mumbai/India—A Complex Adaptive System Perspective," *Habitat International* (2016), http://dx.doi.org/10.1016/j.habitatint.2015.12.017, 5.

47 "of the partition": Aromar Revi, "Lessons from the Deluge," 3912.

47 even the courts: See Ranger et al., "An Assessment of the Potential Impact of Climate Change on Flood Risk in Mumbai," 156.

47 swamped by floodwaters: Carsten Butsch et al. note that while many improvements have been made to Mumbai's warning systems and disaster management practices, "there are also doubts about Mumbai's disaster preparedness. First some of the infrastructures created, are not maintained as good practice would demand; second, many of the measures announced have not been finalized (especially the renovation of the city's water system) and third, informal practices prohibit planning and applying measures." ("Risk Governance in the Megacity Mumbai/India," 9–10).

47 in recent years: Because of emergency measures the death toll of the 2013 Category 5 storm, Cyclone Phailin, was only a few dozen. See the October 14, 2013, CNN report, "Cyclone Phailin: India Relieved at Low Death Toll."

48 planning for disasters: Ranger et al. observe that while Mumbai administration's risk reducing measures are commendable "they do not appear to consider the potential impacts of climate change on the long-term planning horizon." ("An Assessment of the Potential Impact of Climate Change on Flood Risk in Mumbai," 156).

48 "post-disaster response": Friedemann Wenzel et al., "Megacities— Megarisks," *Natural Hazards* 42 (2007): 481–91, 486.

48 disasters of this kind: The Municipal Corporation of Great Mumbai's booklet *Standard Operating Procedures for Disaster Management Control* (available at http://www.mcgm.gov.in/irj/portalapps/com.mcgm.aDisasterMgmt/docs/MCGM_SOP.pdf) is

explicitly focused on floods and makes no mention of cyclones. Cyclones are mentioned only generically in the Municipal Corporation's 2010 publication *Disaster Risk Management Master Plan: Legal and Institutional Arrangements; Disaster Risk Management in Greater Mumbai*, and that too mainly in the context of directives issued by the National Disaster Management Authority, which was established by the country's Disaster Management Act of 2005. The *Maharashtra State Disaster Management Plan* (draft copy) is far more specific, and it includes a lengthy section on cyclones (section 10.4) and the following recommendation: "Evacuate people from unsafe buildings/structures and shift them to relief camps/sites." However, its primary focus is on rural areas, and it does not make any reference (probably for jurisdictional reasons) to a possible evacuation of Mumbai (the plan is available here: http://gadchiroli. nic.in/pdf-files/state-disaster.pdf). The *Greater Mumbai Disaster Management Action Plan: Risk Assessment and Response Plan*, vol. 1, does recognize the threat of cyclones, and even lists the areas that may need to be evacuated (section 2.8). But this list accounts for only a small part of the city's population; the plan does not provide for the possibility that an evacuation on a much larger scale, involving most of the city's people, may be necessary. The plan is available here: http://www.mcgm.gov.in/irj/portalapps/com. mcgm.aDisasterMgmt/docs/Volume%201%20(Final).pdf.

48 projects are located: According to an article published in the *Indian Express* on April 30, 2015, "60 sea-front projects, mostly super luxury residences," were waiting for clearance "along Mumbai's western shoreline." http://indianexpress.com/article/cities/mumbai/govt-forms-new-panel-fresh-hope-for-117-stalled-crz-projects/. The Maharashtra government is also opening many unbuilt sea-facing areas, like the city's old salt pans, to construction (see *The Hindu's Business Line* of August 22, 2015: http://m.thehindubusinessline. com/news/national/salt-pan-lands-in-mumbai-to-be-used-for-development-projects/article7569641.ece).

49 corrugated iron: Carsten Butsch et al., "Risk Governance in the Megacity Mumbai/India," 5.

50 Arabian Sea: Cf. C. W. B. Normand, *Storm Tracks in the Arabian Sea*,

India Meteorological Department, 1926. I am grateful to Adam
Sobel for this reference.

51   city as well: During the 2005 deluge "The waterlogging lasted for
over seven days in parts of the suburbs and the flood water level
had risen by some feet in many built-up areas." B. Arunachalam,
"Drainage Problems of Brihan Mumbai," 3909.

51   illness and disease: See Carsten Butsch et al., "Risk Governance in
the Megacity Mumbai/India," 4.

51   forty thousand beds: Cf. Municipal Corporation of Greater
Mumbai's *City Development Plan*, section on "Health" (9.1; available
here: http://www.mcgm.gov.in/irj/go/km/docs/documents/
MCGM%20Department%20List/City%20Engineer/Deputy%20
City%20Engineer%20(Planning%20and%20Design)/City%20
Development%20Plan/Health.pdf).

51   urban limits: Aromar Revi, "Lessons from the Deluge," 3912.

51   rising seas: Natalie Kopytko, "Uncertain Seas, Uncertain Future for
Nuclear Power," *Bulletin of the Atomic Scientists* 71, no. 2 (2015): 29–38.

52   "safety risks": Ibid., 30–31.

52   models predict: "All the models are indicating an increase in mean
annual rainfall as compared to the observed reference mean of
1936 mm, and the average of all the models in 2350 mm [by 2071–
2099]." Arun Rana et al., "Impact of Climate Change on Rainfall
over Mumbai using Distribution-Based Scaling of Global Climate
Model Projections," *Journal of Hydrology: Regional Studies* 1 (2014):
107–28, 118. See also Dim Coumou and Stefan Rahmstorf, "A Decade
of Weather Extremes," *Nature Climate Change* 2 (July 2012): 491–96:
"Many lines of evidence . . . strongly indicate that some types of
extreme weather event, most notably heatwaves and precipitation
extremes, will greatly increase in a warming climate and have
already done so" (494).

53   become uninhabitable: Aromar Revi notes: "There is a clear need to
rationalize land cover and land use in Greater Mumbai in keeping
with rational ecological and equitable economic considerations. . . .
The key concern here is that developers' interests do not
overpower 'public interest,' that the rights of the poor are upheld;

else displacement from one location will force them to relocate to another, often more risk-prone location" ("Lessons from the Deluge," 3914).

53 threatened neighborhoods: *Climate Risks and Adaptation in Asian Coastal Megacities: A Synthesis Report,* World Bank, 2010 (available at file:///C:/Users/chres/Desktop/Current/research/coastal_ megacities_fullreport.pdf). The report includes a ward-by-ward listing of the areas of Kolkata that are most vulnerable to climate change (88).

55 "below this point": http://www.nytimes.com/2011/04/21/world/ asia/21stones.html.

56 with the "sublime": Cf. William Cronon, "The Trouble with Wilderness; or Getting Back to the Wrong Nature," in *Uncommon Ground: Rethinking the Human Place in Nature,* ed. William Cronon (New York: W. W. Norton, 1995), 69–90: "By the second half of the nineteenth century, the terrible awe that Wordsworth and Thoreau regarded as the appropriately pious stance to adopt in the presence of their mountaintop God was giving way to a much more comfortable, almost sentimental demeanor" (6).

57 they had caused: Cf. A. K. Sen Sarma, "Henry Piddington (1797– 1858): A Bicentennial Tribute," in *Weather* 52, no. 6 (1997): 187–93.

57 "five to fifteen feet": Henry Piddington, "A letter to the most noble James Andrew, Marquis of Dalhousie, Governor-General of India, on the storm wave of the cyclones in the Bay of Bengal and their effects in the Sunderbunds, Baptist Mission Press" (Calcutta, 1853). Quoted in A. K. Sen Sarma, "Henry Piddington (1797–1858): A Bicentennial Tribute."

60 "stretches of farmland": Adwaita Mallabarman, *A River Called Titash,* tr. Kalpana Bardhan (Berkeley: University of California Press, 1993), 16–17.

60 "geography books": Ibid., 12.

61 "Flower-Fruit Mountain": *The Journey to the West,* tr. and ed. Anthony C. Yu (Chicago: University of Chicago Press, 1977.

62 "had to be evacuated": David Lipset, "Place in the Anthropocene: A Mangrove Lagoon in Papua New Guinea in the Time of Rising Sea-

Levels," *Hau: Journal of Ethnographic Theory* 4, no. 3 (2014): 215–43, 233.

63 "inhuman nature": Henry David Thoreau, *In the Maine Woods* (1864).

64 likes and dislikes: Julie Cruikshank, *Do Glaciers Listen? Local Knowledge, Colonial Encounters and Social Imagination* (Vancouver: University of British Columbia Press, 2005), 8.

64 "living and non-living": Julia Adeney Thomas, "The Japanese Critique of History's Suppression of Nature," *Historical Consciousness, Historiography and Modern Japanese Values*, International Symposium in North America, International Research Center for Japanese Studies, Kyoto, Japan, 2002, 234.

64 "never saw an ape": Quoted by Giorgio Agamben in *The Open: Man and Animal*, tr. Kevin Attell (Stanford, CA: Stanford University Press, 2004), Kindle edition, loc. 230.

65 "words and texts": Michael S. Northcott, *A Political Theology of Climate Change* (Cambridge: Wm. B. Eerdmans Publishing, 2013), 34. See also Lynn White, "The Historical Roots of Our Ecological Crisis," *Science* 155 (1967): "Christianity is the most anthropocentric religion the world has seen" (1205).

66 in recorded history: Alexander M. Stoner and Andony Melathopoulos, *Freedom in the Anthropocene: Twentieth-Century Helplessness in the Face of Climate Change* (New York: Palgrave, 2015), 10.

66 "without a Summer": Cf. Michael E. Mann, *The Hockey Stick and the Climate Wars* (New York: Columbia University Press, 2012), 39; Gillen D'Arcy Wood, "1816, the Year without a Summer," *BRANCH: Britain, Representation and Nineteenth-Century History*, ed. Dino Franco Felluga, extension of *Romanticism and Victorianism on the Net* (http://www.branchcollective.org/); and Gillen D'Arcy Wood, *Tambora: The Eruption That Changed the World* (Princeton, NJ: Princeton University Press, 2015).

67 John Polidori: Fiona MacCarthy, *Byron: Life and Legend* (New York: Farrar, Strauss and Giroux, 2002), 292.

67 "amid the darkness": Quoted by Gillen D'Arcy Wood, "1816, the Year without a Summer"; see also John Buxton, *Byron and Shelley: The History of a Friendship* (London: Macmillan, 1968), 10.

67 "August Darvell": Fiona MacCarthy, *Byron*, 292.

67 "vital warmth": Quoted by John Buxton, *Byron and Shelley*, 14.

68 "as we choose": Geoffrey Parker, *Global Crisis*, loc. 17871.

72 "umbrella": Margaret Atwood, *In Other Worlds: SF and the Human Imagination* (New York: Nan A. Talese/Doubleday, 2011).

74 "defend this autonomy": Timothy Mitchell, *Carbon Democracy: Political Power in the Age of Oil* (London: Verso, 2011), Kindle edition, loc. 474.

74 First World War: Ibid., locs. 430, 578.

74 transportation and distribution: Ibid., locs. 680–797: "Whereas the movement of coal tended to follow dendritic networks, with branches at each end but a single main channel, creating potential choke points at several junctures, oil flowed along networks that had the properties of a grid, like an electricity network, where there is more than one possible path and the flow of energy can switch to avoid blockage or overcome breakdowns" (797).

74 from coal to oil: Ibid., loc. 653.

74 "energy flows": Ibid., loc. 645.

74 substance itself is not: Stephanie LeMenager's apt summation in *Living Oil*: "Oil has been shit and sex, the essence of entertainment" (92).

75 Sebastião Salgado: There are, however, many exceptions. For a full account, see the chapter "The Aesthetics of Petroleum" in Stephanie LeMenager's *Living Oil*.

76 "literally inconceivable": The piece is reprinted in the nonfiction collections published under the titles *The Imam and the Indian* (New Delhi: Penguin India, 2002) and *Incendiary Circumstances* (Boston: Houghton Mifflin, 2004).

77 "historical chronicle": Leo Tolstoy, "A Few Words Apropos of the Book *War and Peace*."

77 preceding forms: Donna Tussing Orwin, "Introduction," in *Tolstoy on War: Narrative Art and Historical Truth in "War and Peace,"* ed. Rick McPeak and Donna Tussing Orwin (Ithaca, NY: Cornell University Press, 2012), 3.

78 "being an egotist": *Eight Letters from Charlotte Brontë to George Henry Lewes*, November 1847–October 1850: http://www.bl.uk/collection-

items/eight-letters-from-charlotte-bront-to-george-henry-lewes-
november-1847-october-1850.

78   "collective metamorphosis": Rob Nixon, *Slow Violence and the
Environmentalism of the Poor* (Cambridge, MA: Harvard University
Press, 2011), 87–88.

79   "Great Acceleration": Cf. Will Steffen, Jacques Grinevald, et al.,
"The Anthropocene: Conceptual and Historical Perspectives,"
Philosophical Transactions of the Royal Society 369 (2011): 842–67.

79   "produce isolation": Guy Debord, *The Society of the Spectacle*, 3rd ed.,
tr. Donald Nicholson-Smith (New York: Zone Books, 1994), thesis 28.

79   "as progress": Bruno Latour, *We Have Never Been Modern*, tr.
Catherine Porter (Cambridge, MA: Harvard University Press, 1993),
Kindle edition, loc. 1412.

79   powerful presence: As Latour notes, "the word 'modern' is always
being thrown into the middle of a fight, in a quarrel where there are
winners and losers." Ibid., loc. 269.

79   "used-up" after all: As John Barth once suggested in "The Literature
of Exhaustion," in *The Friday Book: Essays and Other Nonfiction*
(Baltimore: John Hopkins University Press, 1984).

80   avant la lettre: Thus, for example, the scientist and water expert
Peter Gleick writes, in relation to the drought in California: "But
here is what I fear, said best by John Steinbeck in *East of Eden*: 'And
it never failed that during the dry years the people forgot about the
rich years, and during the wet years they lost all memory of the dry
years. It was always that way.'" (*Learning from Drought: Five Priorities
for California*, February 10, 2014; available here: http://scienceblogs.
com/significantfigures/index.php/2014/02/10/learning-from-
drought-five-priorities-for-california/).

80   global temperatures: John L. Brooke, *Climate Change and the Course of
Global History*, 551.

82   move beyond language: "We need . . . to 'decolonize thought,'
in order to see that thinking is not necessarily circumscribed
by language, the symbolic, or the human." Eduardo Kohn, *How
Forests Think: Toward an Anthropology beyond the Human* (Berkeley:
University of California Press, 2013), Kindle edition, loc. 949. See

also John Zerzan, *Running on Emptiness: The Pathology of Civilization* (Los Angeles: Feral House, 2002), 11: "Language seems often to close an experience, not to help ourselves be open to an experience."

83   has literary fiction: Sergio Fava discusses some of the visual artists who have addressed climate change in his book *Environmental Apocalypse in Science and Art: Designing Nightmares* (London: Routledge, 2013).

84   "the written word": Quoted in Arran E. Gare, *Postmodernism and the Environmental Crisis* (London: Routledge, 1995), 21.

84   "world more unlivable": The words are Franco Moretti's from *The Bourgeois* (London: Verso, 2013), 89.

PART II

88   half a million people: "7 Places Forever Changed by Eco-Disasters," http://www.mnn.com/earth-matters/wilderness-resources/ photos/7-places-forever-changed-by-eco-disasters/bhola-island. See also George Monbiot, *Heat: How to Stop the Planet from Burning* (Cambridge, MA: South End Press, 2007), 21.

88   90 percent were women: Varsha Joshi, "Climate Change in South Asia: Gender and Health Concerns," in *Climate Change: An Asian Perspective*, ed. Surjit Singh et al. (Jaipur: Rawat Publications, 2012), 209–26, 213.

88   all along the coastline: Anwar Ali, "Impacts of Climate Change on Tropical Cyclones and Storm Surges in Bangladesh," in *Proceedings of the SAARC Seminar on Climate Variability in the South Asian Region and Its Impacts* (Dhaka: SAARC Meteorological Research Centre, 2003), 130–36, 133. See also M. J .B. Alam and F. Ahmed, "Modeling Climate Change: Perspective and Applications in the Context of Bangladesh," in *Indian Ocean Tropical Cyclones and Climate Change,* ed. Yassine Charabi (London: Springer, 2010), 15–23.

88   the oceans are rising: Cf. "World's River Deltas Sinking Due to Human Activity, Says New Study," http://www.sciencedaily.com/ releases/2009/09/090920204459.htm. See also "Land Subsidence at Aquaculture Facilities in the Yellow River Delta, China," http://

onlinelibrary.wiley.com/doi/10.1002/grl.50758/abstract. In parts of
India, land has subsided by more than thirty feet. Cf. Karen Piper,
The Price of Thirst, loc. 581.

88   groundwater and oil: "Retreating Coastlines," http://www.
straitstimes.com/asia/retreating-coastlines.

89   being especially imperiled: Cf. "South Asia's Sinking Deltas," http://
poleshift.ning.com/profiles/blogs/south-asia-s-sinking-deltas. See
also "InSAR Measurements of Compaction and Subsidence in the
Ganges-Brahmaputra Delta, Bangladesh," http://onlinelibrary.wiley.
com/doi/10.1002/2014JF003117/abstract, and "The Quiet Sinking of
the World's Deltas," http://www.futureearth.org/blog/2014-apr-4/
quiet-sinking-worlds-deltas.

89   acres of agricultural land: Andrew T. Guzman, *Overheated: The
Human Cost of Climate Change* (Oxford: Oxford University Press,
2013), 156.

89   chain, may disappear: P. S. Roy, "Human Dimensions of Climate
Change: Geospatial Perspective," in *Climate Change, Biodiversity, and
Food Security in the South Asian Region*, ed. Neelima Jerath et al. (New
Delhi: Macmillan, 2010), 18–40, 32.

89   75 million in Bangladesh: Pradosh Kishan Nath, "Impact of Climate
Change on Indian Economy: A Critical Review," in *Climate Change:
An Asian Perspective*, ed. Surjit Singh et al., 78–105, 91. For more on
environmental refugees, see also Fred Pearce, *When the Rivers Run
Dry: Water—The Defining Crisis of the Twentieth Century* (Boston:
Beacon Press, 2006), Kindle edition, chap. 4.

89   will be displaced: Carlyle A. Thayer, "Vietnam," in *Climate Change
and National Security: A Country-Level Analysis*, ed. Daniel Moran
(Washington, DC: Georgetown University Press, 2011) 29–41, 30.

89   turning into desert: Lester R. Brown, *World on the Edge: How to
Prevent Environmental and Economic Collapse* (New York: W. W.
Norton, 2011), 40.

89   supply by a quarter: Gwynne Dyer, *Climate Wars*, loc. 987.

89   "only meager crops": Fred Pearce, *When the Rivers Run Dry*, loc. 356.

89   losses of $65 billion: Joanna I. Lewis, "China," in *Climate Change
and National Security: A Country Level Analysis*, ed. Daniel Moran
(Washington, DC: Georgetown University Press), 9–26, 13–14.

See also Kenneth Pomeranz, *Water, Energy, and Politics: Chinese Industrial Revolutions in Global Environmental Perspective* (New York: Bloomsbury, forthcoming), 5.

90   "human race come together": Kenneth Pomeranz, "The Great Himalayan Watershed: Water Shortages, Mega-Projects, and Environmental Politics in China, India, and Southeast Asia," *Revue d'histoire modern et contemporaine* 62, no. 1 (January–March, 2015): 1, 6–47. I am grateful to the author for letting me have an English-language version of this article, published in shorter form in *New Left Review* 58 (2009) and elsewhere.

90   disappear by 2050: Kenneth Pomeranz, "The Great Himalayan Watershed," 32.

90   Indus floods of 2010: Varsha Joshi, "Climate Change in South Asia: Gender and Health Concerns," 209–26, 215, and Pradosh Kishan Nath, "Impact of Climate Change on Indian Economy: A Critical Review," 78–105, 88, both in *Climate Change: An Asian Perspective*, ed. Surjit Singh et al. See also Dewan Abdul Quadir et al., "Climate Change and Its Impacts on Bangladesh Floods over the Past Decades," and Anwar Ali, "Climate Change Impacts and Adaption Assessment in Bangladesh," 165–77, 169, both in *Proceedings of the SAARC Seminar on Climate Variability in the South Asian Region and its Impacts* (Dhaka: SAARC Meteorological Research Centre, 2003). And Wen Stephenson, *What We're Fighting for "Now" Is Each Other: Dispatches from the Front Lines of Climate Justice* (Boston: Beacon Press, 2015), Kindle edition, loc. 391.

90   of them are in Asia: Cf. Johan Rockström et al., "Planetary Boundaries: Exploring the Safe Operating Space for Humanity," *Ecology and Society* 14, no. 2 (2009): 32.

90   disproportionately by women: Surjit Singh, "Mainstreaming Gender in Climate Change Discourse," in *Climate Change: An Asian Perspective*, ed. Surjit Singh et al., 180–208, 184.

90   214 million people: Kenneth Pomeranz, "The Great Himalayan Watershed," 6–47, 7.

91   late in the twentieth century: Thus, as of 2015, the per capita carbon dioxide emissions of the United States and Germany, measured in metric tons, was 17.6 and 9.1, respectively, while the same figures for

China and India were 6.2 and 1.7, respectively. See "World Bank: CO2 Emissions (Metric Tons Per Capita)," http://data.worldbank.org/indicator/EN.ATM.CO2E.PC/countries.

91 back to the 1930s: Spencer R. Weart, *The Discovery of Global Warming* (Cambridge, MA: Harvard University Press, 2003), 1–2.

91 Mauna Loa Observatory in Hawaii: Charles D. Keeling, "Rewards and Penalties of Monitoring the Earth," *Annual Review of Energy and the Environment* 23 (1998): 25–82, 39–42.

92 in the late 1980s: Cf. "The History of Carbon Dioxide Emissions," http://www.wri.org/blog/2014/05/history-carbon-dioxide-emissions. The Four Tigers were South Korea, Taiwan, Hong Kong, and Singapore (John L. Brooke, *Climate Change and the Course of Global History*, 536). They were soon to be followed by the Southeast Asian economies.

92 asphyxiate in the process: Paul G. Harris notes, "If everyone were to live like Americans, the world would require ten times the energy it is using today." See Paul G. Harris, *What's Wrong with Climate Politics and How to Fix It* (Cambridge: Polity Press, 2013), 109.

93 to much of the world: Thus, for example, vulcanologist Bill McGuire cites 1769 CE as a key date in the history of the Anthropocene because that was the year when Richard Arkwright invented the spinning jenny, a machine that would serve as a critical link in the transition to carbon-intensive forms of production: "Arkwright's legacy," writes McGuire, "is nothing less than the industrialization of the world." See Bill McGuire, *Waking the Giant*, Kindle edition, loc. 363. For Timothy Morton, on the other hand, the key moment is April 1784, a date about which, he asserts, "we can be uncannily precise" because that was when James Watt "patented the steam engine." See Timothy Morton, *Hyperobjects*, loc. 210.

93 "particularly in the U.S.": Anil Agarwal and Sunita Narain, *Global Warming in an Unequal World: A Case of Environmental Colonialism* (New Delhi: Centre for Science and Environment, 1991), 1.

94 removed from each other: These connections and processes are explored at length by Jack Goody in *The Eurasian Miracle* (Cambridge: Polity Press, 2010).

94 and the Indian subcontinent: Sheldon Pollock, *The Language of the*

Gods in the World of Men: Sanskrit, Culture, and Power in Premodern India (Berkeley: University of California Press, 2009) 437–52.

94 by the Islamic expansion: Ibid., 489–94.

94 across the Eurasian landmass: Cf. John L. Brooke, *Climate Change and the Course of Global History*, 413, 418.

94 parts of the planet: Cf. Geoffery Parker, *Global Crisis*, loc. 17565: "The return of a warmer climate [in the eighteenth century] had broken the 'fatal synergy'"; and particularly John L. Brooke, *Climate Change and the Course of Global History*, 413–67.

94 Middle East, and India: Cf. Richard M. Eaton and Philip S. Wagoner, "Warfare on the Deccan Plateau, 1450–1600: A Military Revolution in Early Modern India?" *Journal of World History* 25, no. 1 (March 2014): 5–50.

94 savants from elsewhere: Cf. Prasannan Parthasarathi, *Why Europe Grew Rich and Asia Did Not: Global Economic Divergence, 1600–1850* (Cambridge: Cambridge University, Press, 2011), 191, and Richard Grove, "The Transfer of Botanical Knowledge between Europe and Asia, 1498–1800," *Journal of the Japan-Netherlands Institute* 3 (1991): 160–76.

94 "250 years" . . . by Jesuits: George Gheverghese Joseph, *The Crest of the Peacock: Non-European Roots of Mathematics*, 3rd ed. (Princeton, NJ: Princeton University Press, 2011), 439.

95 "stored up in the East": Jonardon Ganeri, *Indian Logic: A Reader* (London: Routledge, 2001), 7.

95 "ten years after his death": Jonardon Ganeri, "Philosophical Modernities: Polycentricity and Early Modernity in India," *Royal Institute of Philosophy Supplement* 74 (2014): 75–94, 87.

95 "to Europe and back": Jonardon Ganeri, "Philosophical Modernities," 86.

95 rest of the world: Sanjay Subrahmanyam, "Hearing Voices: Vignettes of Early Modernity in South Asia, 1400–1750," *Daedalus* 127, no. 3 (1998): 75–104.

95 its own uniqueness: For more on this, see Jack Goody, *The Theft of History* (Cambridge: Cambridge University Pres, 2006).

96 "medieval economic revolution": This episode may itself have had a complex relationship with climatic variations. Cf. Mark Elvin,

The Retreat of the Elephants: An Environmental History of China (New Haven, CT: Yale University Press, 2004) 6, 56.

96 Yellow, and Yangtze Rivers: Ibid., 23.

97 off to visit it: Quoted in ibid., 20–21.

97 easily accessible locations: Kenneth Pomeranz, *The Great Divergence: China, Europe, and the Making of the Modern World Economy* (Princeton, NJ: Princeton University Press, 2000), 46.

98 no cause for astonishment: Quoted by Mark Elvin, *The Retreat of the Elephants*, 68–69.

98 "vigor and virtuosity": Ibid., 69.

100 upon its surface: Amitav Ghosh, *The Glass Palace* (New Delhi: Penguin India, 2000), 130–31.

100 a millennium or more: Marilyn V. Longmuir, *Oil in Burma: The Extraction of "Earth-Oil" to 1914* (Banglamung, Thailand: White Lotus Press, 2001). I am grateful to Dr. Rupert Arrowsmith for bringing Longmuir's book to my notice. See also Khin Maung Gyi, *Memoirs of the Oil Industry in Burma, 905 A.D.–1980 A.D.* (1989).

100 "largest in the world": Marilyn V. Longmuir, *Oil in Burma*, 8.

101 "along the bank": Quoted in ibid., 9–10.

101 by the Burmese: Ibid., 24.

101 forty-six thousand barrels: Ibid., 46.

102 120 oil wells: Thant Myint-U, *The Making of Modern Burma* (Cambridge: Cambridge University Press, 2001), 181.

102 France and England: Ibid., 112–15.

102 1850s onward: Ibid., 149.

103 "Titusville, Pennsylvania": Marilyn V. Longmuir, *Oil in Burma*, 7.

103 Hudson River in 1807: Cf. Blair B. Kling, *Partner in Empire: Dwarkanath Tagore and the Age of Enterprise in Eastern India* (Berkeley: University of California Press, 1977), 65. According to Kling, the first steam engine to reach India was sent from Birmingham to Calcutta in 1817 or 1818; it was bought by the governing authority.

104 after the Netherlands: Parthasarathi, *Why Europe Grew Rich and Asia Did Not*, 229.

104 "flotilla in service": Henry T. Bernstein, *Steamboats on the Ganges* (Calcutta: Orient Longmans, 1960), quoted in Saroj Ghose, "Technology: What Is It?," in *Science, Technology, Imperialism, and*

War, ed. Jyoti Bhusan Das Gupta (New Delhi: Pearson, 2007), 197–
260, 233.

104 forty-three thousand pounds sterling: Arnold van Beverhoudt,
*These Are the Voyages: A History of Ships, Aircraft, and Spacecraft Named
Enterprise* (self-published, 1990), 52.

104 nature of the journey: Ibid.

105 soot and cinders: Amitav Ghosh, *Flood of Fire* (New York: Farrar,
Straus and Giroux).

105 commercial infrastructure: Blair B. Kling, *Partner in Empire*, 61.

106 coal in Bengal: Prasannan Parthasarathi, *Why Europe Grew Rich and
Asia Did Not*, 231.

106 and the United States: Ibid., 233.

106 accessing British ports: R. A. Wadia, *The Bombay Dockyard and the
Wadia Master Builders* (Bombay, 1955), 126–27, quoted in Saroj Ghose,
"Technology: What Is It?," 225.

106 ships and sailors ("lascars"): See Prasannan Parthasarathi, *Why
Europe Grew Rich and Asia Did Not*, 211.

106 "years put together": Satpal Sangwan, "The Sinking Ships: Colonial
Policy and the Decline of Indian Shipping, 1735–1835," in *Technology
and the Raj: Western Technology and Technical Transfers to India, 1700–
1947*, ed. Rory MacLeod and Deepak Kumar (New Delhi, 1995), 137–
52, quoted by Saroj Ghose, "Technology: What Is It?," 225.

107 "my own building": Quoted by Anne Bulley in *The Bombay Country
Ships, 1790–1833* (Richmond: Curzon Press, 2000), 246.

107 "not being imitated": Timothy Mitchell, *Carbon Democracy*, loc. 404.

108 competitors elsewhere: See Prasannan Parthasarathi, *Why Europe
Grew Rich and Asia Did Not*, 225 and 244: "The absence of state
support for industrial development in India stands in stark contrast
to the policies found in Europe in the eighteenth and nineteenth
centuries."

109 critical to its advancement: Cf. ibid., 258–63.

109 "European imperial powers": Dipesh Chakrabarty, "Climate and
Capital: On Conjoined Histories," *Critical Inquiry* 41 (Autumn 2014): 15.

111 "inevitable doom": David Archer, *The Long Thaw: How Humans Are
Changing the Next 100,000 Years of Earth's Climate* (Princeton, NJ:
Princeton University Press, 2009), 172.

111   "bare like locusts": *Young India*, December 20, 1928, 422.

112   "insatiable desires": I owe this reference to Liang Yongjia, Prasenjit Duara, and Tansen Sen; my thanks to all of them for their help.

113   "and nation-building": Prasenjit Duara, *The Crisis of Global Modernity: Asian Traditions and a Sustainable Future* (Cambridge: Cambridge University Press, 2015), 236.

113   wasteful of resources: Cf. Kaoru Sugihara, "East Asian Path," *Economic and Political Weekly* 39, no. 34 (2004): 3855–58.

113   "the Japanese people": Julia Adeney Thomas, "The Japanese Critique of History's Suppression of Nature," *Historical Consciousness, Historiography and Modern Japanese Values*, International Symposium in North America, International Research Center for Japanese Studies, Kyoto, Japan, 2002, 234; my italics.

113   "collapsed around them": Cf. A. Walter Dorn, "U Thant: Buddhism in Action," in *The UN Secretary-General and Moral Authority: Ethics and Religion in International Leadership*, ed. Kent J. Kille (Washington, DC: Georgetown University Press, 2007), 143–86. This article is also available as a pdf at http://walterdorn.net/pdf/UThant-BuddhismInAction_Dorn_SG-MoralAuthority_2007.pdf.

115   beings as a species: Dipesh Chakrabarty, "The Human Condition in the Anthropocene," The Tanner Lectures on Human Values, Yale University, 2015.

115   "of the present": Watsuji Tetsuro, *A Climate: A Philosophical Study*, tr. Geoffrey Bownas (Ministry of Education, Printing Bureau, Japanese Government, 1961), 30. I am grateful to Giorgio Agamben for bringing this work to my attention.

119   "or human-made systems": Dipesh Chakrabarty, "The Climate of History: Four Theses," *Critical Inquiry* 35 (Winter 2009): 208.

119   through human agency: Cf. Julia Adeney Thomas, "The Present Climate of Economics and History," in *Economic Development and Environmental History in the Anthropocene: Perspectives on Asia and Africa*, ed. Gareth Austin (London: Bloomsbury Academic, forthcoming), 4.

120 favored by the USSR: Cf. Frances Stonor Saunders, "Modern Art Was CIA 'Weapon,'" *The Independent*, October 21, 1995. See also Joel Whitney, *FINKS: How the CIA Tricked the World's Best Writers*, (London: OR Books, 2016), chap. 2.

120 "and the artist": Roger Shattuck, *The Banquet Years: The Origins of the Avant-Garde in France, 1885 to World War I* (New York: Vintage, 1968), 326.

120 passion for dams: Cf. Kenneth Pomeranz, "The Great Himalayan Watershed: Water Shortages, Mega-Projects, and Environmental Politics in China, India, and Southeast Asia," 19 (published in French as "Les eaux de l'Himalaya: Barrages géants et risques environnementaux en Asia contemporaine," in *Revue d'histoire modern et contemporaine* 62, no. 1 [January–March 2015]: 6–47); for Mao's "War against Nature," see Judith Shapiro, *Mao's War against Nature: Politics and the Environment in Revolutionary China* (Cambridge: Cambridge University Press, 2001).

120 "world [they] depict": Franco Moretti, *The Bourgeois*, 89.

121 "the official order": Arran E. Gare, *Postmodernism and the Environmental Crisis* (London: Routledge, 1995), 16.

121 perspective of the Anthropocene: As Stephanie LeMenager points out, even Upton Sinclair, a committed socialist and "one of the most ideologically driven American novelists," ends up romanticizing the gasoline-powered culture of cars. See Stephanie LeMenager, *Living Oil*, 69.

121 "becomes a commodity": Guy Debord, *The Society of the Spectacle*, 3rd ed. (New York: Zone Books, 1994), 59.

123 "with us always": Roger Shattuck, *The Banquet Years*, 25.

123 predecessor obsolete: Bruno Latour, *We Have Never Been Modern*, loc. 1412.

123 "wrong side of history": http://www.slate.com/blogs/lexicon_valley/2014/04/17/the_phrase_the_wrong_side_of_history_around_for_more_than_a_century_is_getting.html.

126 vulnerable to climate change: Cf. *IPCC's Fifth Assessment Report: What's in It for South Asia, Executive Summary* (available at: http://cdkn.org/wp-content/uploads/2014/04/CDKN-IPCC-Whats-in-it-for-South-Asia-AR5.pdf).

127　implicit in it: Friedrich Nietzsche, *On the Genealogy of Morals* (1887), tr. Carol Diethe, ed. Keith Ansell-Pearson (Cambridge: Cambridge University Press, 2006).

127　journey of self-discovery: "This emphasis on individual conscience and its capacity to 'trump' all other arguments in fact seems to be a defining feature of much of what has passed for radical (i.e., revolutionary) politics in the United States since the 1960s. . . . There is a pervasive stress on what each and every individual feels and experiences as providing the ultimate standard of legitimacy, action, and definition of collective goals." See Adam B. Seligman, Robert P. Weller, Michael J. Puett, and Bennett Simon, *Ritual and Its Consequences: An Essay on the Limits of Sincerity* (New York: Oxford University Press, 2008), Kindle edition, loc. 1946.

128　"his own life": http://electricliterature.com/knausgaard-and-the-meaning-of-fiction-2/.

128　"Puritan religiosity": Adam B. Seligman et al., *Ritual and Its Consequences*, loc. 1526.

128　imagining of possibilities: I follow here the notion of the subjunctive that is employed by Adam Seligman and his coauthors in *Ritual and Its Consequences*.

129　"ice shelf broke up": The question "Where were you at 400 ppm?" is posed by Joshua P. Howe in "This Is Nature; This Is Un-Nature: Reading the Keeling Curve," *Environmental History* 20, no. 2 (2015): 286–93, 290.

130　"implement their demands": Ingolfur Blühdorn, "Sustaining the Unsustainable: Symbolic Politics and the Politics of Simulation, *Environmental Politics* 16, no. 2 (2007): 251–75, 264–65.

130　after the First World War: Timothy Mitchell, *Carbon Democracy*, loc. 2998.

130　"They only consume": Roy Scranton, *Learning to Die in the Anthropocene: Reflections on the End of a Civilization* (San Francisco: City Lights Books, 2015), Kindle edition, loc. 640.

131　"legislation and governance": Ibid.

131　"the modern world": Adam B. Seligman et al., *Ritual and Its Consequences*, loc. 171.

131 "mere representation": Guy Debord, *The Society of the Spectacle*, thesis 1.

131 "reestablishes its rule": Ibid, thesis 18; my italics.

132 "moral issue": Naomi Klein makes a powerful case for enframing climate change as a moral issue in her magisterial *This Changes Everything: Capitalism vs. the Climate* (New York: Knopf, 2014). See also the following interview with Michael Mann: http://paulharrisonline.blogspot.in/2015/07/michael-mann-on-climate-change.html.

133 opposite side: I am following here the use of the word *sincerity* in Adam B. Seligman et al., *Ritual and Its Consequences*.

133 acted upon: As Rachel Dyer notes, "all the stuff about changing the light-bulbs and driving less, although it is useful for raising consciousness and gives people some sense of control over their fate, is practically irrelevant to the outcome of this crisis." See Rachel Dyer, *Climate Wars*, loc. 118.

134 "at the same time!": John Maynard Keynes, *The End of Laissez-Faire* (1926).

134 parts of a whole: In the words of Naomi Oreskes and Eric Conway, this is a "quasi-religious faith, hence the label *market fundamentalism*" (*The Collapse of Western Civilization: A View from the Future* [New York: Columbia University Press, 2014], 37).

135 "wicked problem": In one definition "wicked problems are essentially unique, have no definitive formulation, and can be considered symptoms of yet other problems" (Mike Hulme, *Why We Disagree about Climate Change: Understanding Controversy, Inaction and Opportunity* [Cambridge: Cambridge University Press, 2009], 334).

136 last two centuries: This is how Tim Flannery puts it: "America and Australia were created on the frontier, and the citizens of both nations hold deep beliefs about the benefits of endless growth and expansion." See *The Weather Makers: How Man Is Changing the Climate and What It Means for Life on Earth* (New York: Atlantic Monthly Press, 2006), 237.

136 throughout the Anglosphere: That the phenomenon of climate denial has a special place within the Anglosphere is recognized

by many; see, for example, this conversation between George
Monbiot and George Marshall: https://www.youtube.com/
watch?v=ocCCanfgZ4A (at 29 min.). See also "The Strange
Relationship between Global Warming Denial and . . . Speaking
English," *Mother Jones*, http://www.motherjones.com/
environment/2014/07/climate-denial-us-uk-australia-canada-
english. The survey on which the article is based is available here:
http://www.ipsosglobaltrends.com/environment.html. In most
industrialized European countries, by contrast, there is very little
denial, either at the popular or official levels: cf. Elizabeth Kolbert,
"Pieter van Geel, the Dutch environment secretary, described
the European outlook to me as follows: 'We cannot say, "Well, we
have our wealth, based on the use of fossil fuels for the last three
hundred years, and, now that your countries are growing, you may
not grow at this rate, because we have a climate change problem."'"
See chap. 8 of Kolbert's *Field Notes from a Catastrophe: Man, Nature,
and Climate Change* (New York: Bloomsbury, 2006).

136   in the United States: Anthony Giddens notes, "In no other
country is opinion about climate change so acutely divided as in
the US today." See Giddens's *The Politics of Climate Change*, 2nd ed.
(Cambridge: Polity Press, Cambridge, 2011), 89.

137   politics of self-definition: See Michael Shellenberger and Ted
Norhaus, "The Death of Environmentalism": "Environmentalists are
in a culture war whether we like it or not" (10). Similarly Andrew
J. Hoffman, notes, "The debate over climate change in the United
States (and elsewhere) is not about carbon dioxide and greenhouse
gas models; it is about opposing cultural values and worldviews
through which that science is seen" (*How Culture Shapes the Climate
Change Debate* [Stanford, CA: Stanford University Press, 2015],
Kindle edition, loc. 139).

137   to be American: Raymond S. Bradley, *Global Warming and Political
Intimidation: How Politicians Cracked Down on Scientists as the Earth
Heated Up* (Boston: University of Massachusetts Press, 2011), 128.

137   communism, and so on: Cf. George Marshall, *Don't Even Think about
It*, 37: "As Rush Limbaugh says, climate science 'has become a home
for displaced socialists and communists.'"

NOTES TO PART THREE

137   the Cold War: Naomi Oreskes and Erik M. Conway, *Merchants of Doubt: How a Handful of Scientists Obscured the Truth on Issues from Tobacco Smoke to Global Warming* (New York: Bloomsbury, 2010), 214.

137   intimidation: Michael Mann describes his battles with climate change deniers at length in his book *The Hockey Stick and the Climate Wars* (New York: Columbia University Press, 2012). See also Raymond S. Bradley, *Global Warming and Political Intimidation*, 125 and 145–48.

137   energy billionaires: Oreskes and Conway, *Merchants of Doubt*.

137   within the electorate: Elizabeth Kolbert in chap. 8 of *Field Notes from a Catastrophe* names some of the lobbying groups, such as the "Global Climate Coalition, a group that was sponsored by, among others, Chevron, Exxon, Ford, General Motors, Mobil, Shell, and Texaco." See also Tim Flannery, *The Weather Makers*, 239.

137   climate scientists: In any case, as Kevin Lister notes, "Even the most ardent papers on climate change such as the Guardian and Independent continue to devote far more space to advertising high carbon holidays abroad and reporting the most intricate details of Formula 1 than they do on reporting climate change" (*The Vortex of Violence and Why We Are Losing the Battle on Climate Change* [CreateSpace Independent Publishing Platform, 2014]), 21.

138   politics of the Anglosphere: "Denial" is not a major factor in most of the world, as Anthony Giddens notes in *The Politics of Climate Change*: "Surveys taken on a global level show that people in the developing countries are the most concerned about climate change. A cross-cultural study of nine developed and developing countries indicated that about 60 per cent of people interviewed about climate change in China, India, Mexico and Brazil felt a 'high level of concern.'" (104).

138   money and manipulation: For more on this, see Joshua P. Howe's review of Oreskes and Conway, *Merchants of Doubt*, in his article "The Stories We Tell," *Historical Studies in the Natural Sciences* 42, no. 3 (June 2012): 244–54, esp. 253.

139   "of the solution": George Marshall, *Don't Even Think about It*, 75–76. Gwynne Dyer in *Climate Wars* notes that the "US Army War College sponsored a two-day conference on 'The National Security

Implications of Climate Change' in 2007" (loc. 250).

139 "security environment": "Admiral Locklear: Climate Change the Biggest Long-Term Security Risk in the Pacific Region," http://climateandsecurity.org/2013/03/12/admiral-locklear-climate-change-the-biggest-long-term-security-threat-in-the-pacific-region/.

139 "Dept. of Defense": https://www.youtube.com/watch?v=ckjY-FW7-dc.

139 the public sphere: Sanjay Chaturvedi and Timothy Doyle, *Climate Terror: A Critical Geopolitics of Climate Change* (Basingstoke: Palgrave Macmillan, 2015), Kindle edition, locs. 3193–215.

139 "neo-securities are one": Ibid., loc. 3256.

140 climate change: See, for example, Kurt Campbell et al., "The Age of Consequences: The Foreign Policy and National Security Implications of Global Climate Change" (Center for New American Security, 2007). Gwynne Dyer in *Climate Wars* notes of this study that "the lead authors ... include John Podesta, who served as chief of staff to President Clinton in 1998–2000; Leon Fuerth, national security advisor to Vice President Gore ... and R. James Woolsey, Jr., head of the Central Intelligence Agency 1993–95" (loc. 304).

140 "disobedience, and vandalism": Quoted in Roy Scranton, *Learning to Die in the Anthropocene*, loc. 80.

140 a top priority: The FBI, for instance, has named "animal rights extremists and eco-terrorism" as its "highest domestic terrorism priority." See Will Potter, *Green Is the New Red: An Insider's Account of a Social Movement under Siege* (San Francisco: City Lights Books, 2011), 25, 44.

140 post-9/11 era: Giorgio Agamben, *State of Exception*, tr. Kevin Attell (Chicago: University of Chicago Press, 2005), Kindle edition, loc. 40.

140 recent years: See Nafeez Ahmed, "Pentagon Bracing for Public Dissent Over Climate and Energy Shocks, *The Guardian*, June 14, 2013 (also available at http://www.theguardian.com/environment/earth-insight/2013/jun/14/climate-change-energy-shocks-nsa-prism).

140 many different kinds: Cf. Adam Federman, "We're Being Watched: How Corporations and Law Enforcement Are Spying on Environmentalists," *Earth Island Journal* (Summer 2013), also

available at http://www.earthisland.org/journal/index.php/eij/
article/we_are_being_watched/. The term "gray intelligence" was
coined by Dr. Bob Hoogenboom, a Dutch professor of Forensic
Business Studies.

140 "and its impacts": Cf. "Be Prepared: Climate Change Security
and Australia's Defense Force," Climate Council: http://www.
climatecouncil.org.au/uploads/fa8b3c7d4c6477720434d6d10897af18.
pdf.

141 approach to climate change: As Roy Scranton points out, "President
Obama's 2010 *National Security Strategy*, the Pentagon's *2014
Quadrennial Defense Review*, and the Department of Homeland
Security's *2014 Quadrennial Homeland Security Review*, all identify
climate change as a severe and imminent danger" (*Learning to Die in
the Anthropocene*, loc. 86.)

141 are most visible: Lewis R. Gordon, in *What Fanon Said: A
Philosophical Introduction to His Life and Thought* (New York: Fordham
University Press, 2015), writes, "In the colonies the truth was naked,
the 'metropoles' preferred it clothed" (133).

141 of the metropole: "In the colonies the truth was naked, the
'metropoles' preferred it clothed." Cf. ibid.

142 "biopolitics": In Foucault's definition, biopolitics is the "attempt,
starting from the eighteenth century, to rationalize the problems
posed to governmental practice by phenomena characteristic of a
set of living beings forming a population: health, hygiene, birthrate,
life expectancy, race" (Michel Foucault, *The Birth of Biopolitics*, tr.
Graham Burchell [New York: Picador, 2004], 317).

142 "would not exist": Timothy Mitchell, *Carbon Democracy*, loc. 136.

143 political legitimacy: Although some would argue, following
John Rawls, that principles of justice "apply only to the internal
affairs within nations and cannot be extended to apply either to
relations between nations or among all the world's persons" (Steve
Vanderheiden, *Atmospheric Justice: A Political Theory of Climate Change*
[New York: Oxford University Press, 2008], 83).

143 "climate budget": Cf. Tom Athanasiou and Paul Baer, *Dead Heat:
Global Justice and Global Warming* (New York: Seven Stories Press,
2002), 76–85.

143   "anti-immigrant policing": Christian Parenti, *Tropic of Chaos: Climate Change and the New Geography of Violence* (New York: Nation Books, 2012), 225.

143   "combat with the earth": Sanjay Chaturdevi and Timothy Doyle, *Climate Terror*, loc. 2893.

145   military space: Ibid., loc. 2984.

145   mitigatory measures: Thus George Monbiot writes (and I wish it were true): "there is a good political reason for fairness. People are more willing to act if they perceive that everyone else is acting" (*Heat: How to Stop the Planet from Burning* (Cambridge, MA: South End Press, 2007), 43).

146   "90 percent": David Archer, *The Long Thaw: How Humans Are Changing the Next 100,000 Years of Earth's Climate* (Princeton, NJ: Princeton University Press, 2009), 163.

147   accustomed to hardship: Something like this was actually implied by Larry Summers when, as head of the World Bank, he proposed that polluting industries should be relocated to less developed nations: "After all, those living in the Third World couldn't expect to live as long as 'we' do, so what could be wrong with reducing their lifetimes by a miniscule amount . . . ?" See David Palumbo-Liu, *The Deliverance of Others: Reading Literature in a Global Age* (Durham, NC: Duke University Press, 2012), vii–viii. Other economists have applied a similar logic. As George Monbiot points out in *Heat*: "In 1996, for example, a study for the Intergovernmental Panel on Climate Change estimated that a life lost in the poor nations could be priced at $150,00, while a life lost in the rich nations could be assessed at $ 1.5 million" (50).

147   "each food calorie": David Orr, *Down to the Wire*, 33.

148   "fifty-six thousand": James Lawrence Powell, *Rough Winds: Extreme Weather and Climate Change*, Kindle Single, 2011, locs. 212–37.

148   living in isolation: Ibid., loc. 210.

149   "those from the OECD": Cf. Samir Saran and Vivan Sharan, "Unbundling the Coal-Climate Equation," *The Hindu*, October 7, 2015.

149   climate change negotiations: As Clive Hamilton observes, the obsession with growth in developing countries is "perhaps the last

and most potent legacy of colonialism" (*Growth Fetish* [Crow's Nest: Allen & Unwin, 2003], Kindle edition, loc. 232).

150 "ramp of global warming": The phrase is Michael Mann's; cf. http://bigstory.ap.org/article/60abc049a4f14cc3bd2569eac806cbde/noaa-nasa-2015-was-hottest-earth-wide-margin.

150 appeared in December: The texts are, respectively, *Encyclical Letter Laudato Si' of the Holy Father Francis on Care of Our Common Home* (hereafter Encyclical), available at https://laudatosi.com/watch; and *Framework Convention on Climate Change* (hereafter Agreement), published by the United Nations and available at http://unfccc.int/resource/docs/2015/cop21/eng/l09.pdf.

150 are also *texts*: The text of the Agreement is specified as having originally been written in English. No original language is specified in the case of the Encyclical nor is any translator listed, so it must be presumed that the text is at least partly the result of a collaboration.

153 already beyond reach: For an extended discussion of "dangerous limits" and public policy, see Christopher Shaw, *The Two Degree Dangerous Limit: Public Understanding and Decision Making* (London: Routledge, 2015).

153 succeed at scale: Cf. Kevin Anderson, *The Hidden Agenda: How Veiled Techno-Utopias Shore Up the Paris Agreement*, http://kevinanderson.info/blog/the-hidden-agenda-how-veiled-techno-utopias-shore-up-the-paris-agreement/. See also "COP21: Paris Deal Far Too Weak to Prevent Devastating Climate Change, Academics Warn," *The Independent*, January 8, 2016.

154 "experts in technology": Encyclical, 79/106.

154 "technological growth": Ibid., 82/109.

154 "drug addiction," and so on: See, for example, *United Nations Single Convention on Narcotic Drugs* (available at https://www.unodc.org/pdf/convention_1961_en.pdf), resolution 3.

154 "market imperfections": *Kyoto Protocol to the United Nations Framework Convention on Climate Change* (available at http://unfccc.int/resource/docs/convkp/kpeng.pdf), article 2 a/v.

154 "concern for humankind": Agreement, 20.

155 celestial needle: Very soon after the Agreement was reached,

twenty-one climate scientists published a open letter saying that
the deal had succeeded only in kicking "the can down the road
by committing to calculate a new carbon budget for a 1.5 deg C
temperature increase that can be talked about in 2020." See "COP
21: Paris Deal Far Too Weak to Prevent Devastating Climate Change
Academics Warn," *The Independent,* January 8, 2016.

155 "to be human": Encyclical, 10/11.

155 "highest level": Agreement, 17–18/articles 122 and 123.

156 "world's population": Encyclical, 35/50.

157 "interior peace": Ibid., 10/10.

157 *"cry of the poor"*: Ibid., 35/49.

158 "global north and south": Ibid., 36/51.

158 "address climate change": Agreement, 20.

158 "liability or compensation": Ibid., 8/article 52.

159 "human abilities": Encyclical, 16/19.

159 "freedom is limitless": Ibid., 7/6.

159 "but also nature": Ibid., 8/7.

159 politics of climate change: See, for instance, "The Interfaith
Declaration on Climate Change," http://www.interfaithdeclaration.org/.

159 voiced their concern: See "The Hindu Declaration on Climate
Change," http://fore.yale.edu/news/item/hindu-declaration-on-climate-change/; "The Muslim 7-Year Action Plan to Deal with
Climate Change," http://www.arcworld.org/downloads/Muslim-7YP.pdf; and "Global Buddhist Climate Change Collective," http://gbccc.org/who-we-are/.

159 on their own: Timothy Mitchell notes in *Carbon Democracy,*
"existing forms of democratic government appear incapable of
taking the precautions needed to protect the long-term future of
the planet" (loc. 253).

159 group of people: Paul G. Harris addresses this problem at some
length in the chapter entitled "The Cancer of Westphalia: Climate
Diplomacy and the International System" in his book *What's Wrong
with Climate Politics and How to Fix It* (Cambridge: Polity Press, 2013).

161 "from a body": Ruth Irwin, *Heidegger, Politics, and Climate Change:
Risking It All* (New York: Bloomsbury, 2008), 158.